798 K

798 K

The Colorful World of
ANIMALS

The Colorful World of
ANIMALS

Maurice & Jane Burton

LONGMEADOW
PRESS

Contents

First published in the USA by
Longmeadow Press, PO Box 16,
Rowayton Station, Norwalk,
Connecticut 06853

© 1975 Hennerwood
Publications Limited

ISBN 0 904230 06 6

Produced by
Mandarin Publishers Limited
Hong Kong

Printed in Hong Kong

The role of colour

A world without colour would be unimaginable. Yet we tend to take the miracle of natural colour for granted. Man, for all his genius, has never succeeded in reproducing it artificially. Not even the most gifted painter has fully captured the extraordinary splendour of the setting sun, the ever-changing blues, greens and greys of the sea, the subtle shades of a rose in full bloom, the iridescence of a butterfly's wings. But even as we admire nature's beauties we seldom stop to think about the purpose of colour. For the truth is that colour both in flowers and animals is a matter not of accident but of design. The colours of flowers are to attract insects that will pollinate the blossoms, so that seeds will develop to bring forth a new generation of plants. Animals wear colours that are often gayer and more varied than those of the finest blooms; they are not merely decorative, but perform a specific service for the animals concerned.

This book describes and illustrates animals in all kinds of colours and combinations of colours. The pictures are enjoyable for their own sake, but the principal reason for showing them is to provide an opportunity to talk about the purpose of these colours.

'Function' is a word much used in biology today, in the study of plants and animals, and in the study of the working of our own bodies. Fifty years ago the word, derived from a Latin verb meaning to perform, was seldom heard; today it is everyday speech. Colour may function as a means of concealment, particularly in the form of specific camouflage markings. It may be used in courtship displays or aggressive displays, especially in defence of territory. Patches of colour are used more particularly in these two ways and also as signals; so colour may serve as a method of passing information from one member of a species to another, to baffle a pursuer, deceive a predator or indicate a mood.

Let us take a simple example. In Great Britain the most popular bird is the robin. It is a small, brownish bird with a red breast. Indeed, it was first of all known as the redbreast. Centuries ago, people started giving birds the names of people. A black-and-white bird was called Maggie pie (or pied). This later became contracted to magpie. Other similar names are jackdaw and Jenny wren. They are signs of affection.

Wherever English-speaking people went out to other parts of the world they took these names with them and gave them to birds that were similar to those of their native country. In North America there is a bird of the thrush family with red on its front, which is called the American robin. There is an Indian robin, a New Zealand robin, a Peking robin, a Cape robin and twenty-one different Australian robins. In various parts of the world there are bush robins, scrub robins, magpie robins and many others. This only goes to show how the red breast caught people's fancy.

What then is the function of the robin's red breast? In the case of the British robin's breast it is well known. When two rival robins meet face to face one of them will point its beak to the skies, so showing its red breast to the full. Then it slowly sways from one side to the other, as if 'waving a red rag'. If the other robin is intimidated by this it will fly away. If not, it will respond by pointing its own beak upwards and swaying from side to side. Usually one of the two robins, after a while, gives way and flies off. Should the two stand their ground, then a fight takes place.

The experiment has been tried of fixing the red feathers from a dead robin's breast on a wire to a branch. The robin living there comes forward, displays its

Robin The British robin, $5\frac{1}{2}$ ins long, will keep close to any large animal, especially one with hoofs, to seek out insects in the earth which that animal disturbs as it walks. It was a short step for the bird to transfer this habit to people, especially those working in a garden. Because a robin is so fearless, a state of friendship readily springs up between it and its human neighbours. Today the robin is beyond doubt the favourite bird in Britain and the one which, because of its red breast, everyone can recognize instantly.

The red breast is not used in courtship, as is sometimes said, but is employed exclusively for defending a territory. The male, unlike most birds, holds his territory almost throughout the year and relinquishes it for a short time only, in August. The female which also has a red breast, shares his territory during the breeding season, late December to August, but has her own for the rest of the year. Both males and females will display at other robins entering their territories. So territory is clearly important as a means of ensuring an adequate supply of food, and the red breast is the bird's safeguard.

Eyed Lizard This is the largest European lizard. It is nearly 2 feet long, with a tail of about 16 inches in length. It is excellently camouflaged and the distinctive blue 'eyes' give this lizard its popular name. Many animals have 'eye' markings and there are thought to be various reasons for them. Sometimes the spots are effective in dazzling an opponent, sometimes they are a part of camouflage colouring and break up solid areas of colour on the animal and sometimes dark spots conceal the animal's true eyes.

red breast at the bunch of feathers and, because the red bunch does not go away, attacks it mercilessly. Thus, to a robin redbreast the patch of red feathers is a signal, inciting it to fight and to drive an intruder out of its territory.

In North America there are birds known as flickers, which are a kind of woodpecker. A male flicker has a red or a black mark on either side of its face, termed a moustache. In one experiment a false moustache was fixed on the hen of a mated pair. She was then released and returned to her mate. Immediately he saw her, he attacked her unmercifully and continued to do so until she was caught again by the experimenter and the moustache removed. The male tried to resume friendly relations but the hen would have none of it at first. She went for him and he did not retaliate. Indeed, he made no attempt to defend himself even, once her disguise had been removed and he recognized her as his chosen mate. In the flickers, therefore, the colour and shape of the moustache form a recognition signal.

Colour in animals has many functions in addition to the two examples just discussed. In birds, brilliant plumage is used in courtship like the brilliant patches on many male fishes. Colours in special patterns may form a camouflage, hiding the animals from their enemies. There are many ways in which camouflage may be used. And then some animals are dark coloured, usually dark brown to black, because this helps to absorb heat. Others are white, which is an absence of colour. It is seen in polar animals and in the winter coats of northern animals. A white coat, whether of fur or feathers, has always been thought to help prevent loss of heat from the body, although there is some disagreement on this. There are, however, some animals whose colours seem to serve no biological purpose; for example, a white bat, which is active only at night.

Colours can arise from two sources; mostly they are due to pigments, but many result from special structures and are called physical colours. These are found especially in insects. The iridescence of many beetles is not due to pigments but to the structure of the cuticle or horny skin covering the body. It may look smooth to the naked eye but under the microscope is seen to be riddled with minute air-filled cavities. When white light, such as daylight, falls on this cuticle it is broken up or refracted into its constituent wavelengths, just as when a beam of light passes through a prism. Such colours disappear when a liquid that will flow into the cavities is poured onto the cuticle. Once the air spaces in the cuticle are filled the light is not broken up and the iridescence is lost.

One of the best-known and most brilliant of tropical butterflies is the morpho. It is blue with an iridescence or sheen. This colour is not due to a pigment but to rows of ribs on the scales covering the wings. The microscopic air spaces between the ribs that break up or refract the light, scatter it and reflect it. Some birds' feathers have colours caused in the same way, as do some wasps and beetles. Those beetles, the colour of which is caused by minute cavities in the cuticle, lose their colour when they are dead and the cuticle dries out, because in the shrinking the cavities become distorted.

In many insects the surface of the cuticle is studded with small, usually transparent knobs, known as particles. These scatter, reflect and refract the light. If the particles are larger than the wavelength of light, the insect appears white, as in white butterflies. When the particles are smaller than the wavelength of the light they scatter the shorter wavelengths, producing blue, as in many dragonflies.

By now we are beginning to see that the colours of animals are not as simple as, say, colours dabbed on a canvas by an artist using a palette and brushes. Indeed, the more we go into the subject, the more complicated it becomes. Some animals, including many caterpillars and butterflies, derive their pigment from the food they eat. Then there are the many animals that can change colour according to the background they are on. There are others that change colour according to their emotions. An octopus or a cuttlefish can rapidly change from one colour to another; the colours may follow each other like waves in fractions of a second. There are fishes living on coral reefs that will swim through the branches of a coral head and come out the other side a totally different colour. Many fishes lose all colour at night, and turn pale. Others are more brilliantly coloured when dying than they are at any time during life.

Colour changes of this kind are due to pigments and special mechanisms are needed to move the pigments around. These mechanisms are in the special cells

Giraffes at a water hole. Their pattern tends to break up the outline of the body even in bright sunlight, but it is most effective when the animals are standing among trees (see page 102).

Western robin The American robin, 8½ ins long and a very familiar bird in most parts of the United States and Canada, is a member of the thrush family. To a visitor from the United Kingdom it looks and behaves like a British blackbird with a red breast. Its song is a loud, clear carolling and when alarmed it uses a strident chirp that alerts other birds. When feeding, it hops about the lawns in towns and cities searching for worms and insects; and it has a partiality for soft fruits.

In some ways the American robin is like the European robin, for it is a woodland bird that has invaded gardens. It eats insects but also takes more fruit than the original robin. There are other marked differences. It is migratory whereas robin redbreast is a resident, staying more or less in the same place the year round. The red covers the whole breast in contrast to the red throat and upper breast of the European robin. Finally, the red breast of the American robin is used in courtship, and only incidentally in the maintenance of a territory.

Lion The king of the beasts is the one animal that acts as the yardstick for tawny. The sandy colouring is often an excellent camouflage. Lions rarely need protection but the tawny coat helps to hide them when stalking prey. See also page 60.

Tern Seabirds are built and coloured to spend most of their life in the air. Over the sea and against the clouds white seabirds tend to disappear, but they show up clearly against the cliffs. A tern flying over the surface presenting its white front to the water does not disturb fishes swimming near the surface, and so finds them easier to capture. On the shore many of the white or partly-white gulls have the advantage of countershading (see page 116). In other white birds the whiteness is due to the lack of pigment in the plumage (see page 82). There are 39 species of Tern distributed throughout the world and they are found in large numbers in the Pacific regions.

Following page:

Brown bear Brown bears used to be found all over the northern hemisphere. They included grizzlies, Russian bears and other types all much alike, noted for their large size and their strength; they ranged from 6 to 9 ft long and from 450 to 1,690 lb in weight. Few animals would willingly challenge such powerful animals, and armed men do so warily. Bears hardly need the extra protection afforded by camouflage except that they must sometimes rest. That is when they are most vulnerable. Since bears are by nature forest dwellers, a brown coat, toning in both with tree trunks and bare ground, is of some value.

The coat of the brown bear is typically reddish-brown and there is one race, living in Kashmir and the western Himalayas, that is paler than usual and is known as the red or isabelline bear.

All over the northern hemisphere brown bears have been hunted by man almost from the beginning of human history. As a result they have been completely wiped out in places and dwindled in numbers elsewhere. Much of this massacre has been relatively recent and due to the invention of firearms, added to man's fear of this powerful animal. Apart from man, the brown bear has virtually no enemies. Any species of animal with few or no enemies to fear is less constant in colouring than one needing good camouflage to survive. Brown bears are normally a shade of brown but may vary to opposite extremes of cream and black.

in the skin known as chromatophores and are of two kinds. The most common type of chromatophore is found mainly in vertebrates and consists of a cell of very irregular outline with pigment granules inside. It might be compared with a bag of tiny grains of colour, the surface of which extends outwards in branches. The pigment granules are normally concentrated at the centre. They can, however, spread outwards in all directions, so that a cell at one moment colourless can quickly become coloured throughout.

The other kind of chromatophore is found only in invertebrates and consists of a central cell which is normally a small coloured dot. Around it are muscle cells arranged like the spokes of a wheel. When these muscle cells contract they pull on the central dot so that it expands. When they relax the cell contracts again to a mere dot.

The colours in a chromatophore are usually black, red or yellow. In addition, in some animals there are silvery chromatophores. More than one colour may be present in a chromatophore and the differently coloured granules can move independently of each other. This means that a chromatophore can be a colourless cell at one time, green at another, or red; or a mixing of the pigment granules can produce a variety of shades of colours.

How the colour changes is best illustrated by the weed prawns, *Hippolyte*. These are found throughout the world living on seaweed. The best-known species of *Hippolyte* inhabits the coasts of Europe, where it is called the Aesop prawn (Aesop was a hunchback, and the prawn has a hunched back). The Aesop prawn is less than an inch long. When it settles on a red seaweed it goes red. On a brown seaweed it goes brown and on a green seaweed, green. At night it changes to transparent blue no matter what the colour of the seaweed. Apart from the nightly changes it takes about a week to change from one colour to another.

The pigments in the chromatophores of the Aesop prawn are blue, yellow and red. On a red seaweed only the red granules are spread over the whole chromatophore. The others are concentrated at the centre. When the prawn is on a green seaweed the red pigment is withdrawn and the blue and yellow expand. On a brown seaweed the prawn goes liver brown, made up mainly of red pigment granules but with some yellow and blue. At night only the blue granules are spread through the cell; the red and yellow have retired to the centre. Although we cannot be sure what it is that makes the pigment granules move about in this way, we do know what can cause it. Strong light causes the colour change in some animals. In others a raising of the temperature in the surrounding air or water makes them go black. Often it is the kind of light reaching the eye that makes the animal change colour, as in flatfishes like the flounder, sole, turbot and plaice. Some of these fishes can change colour to suit that of the seabed on which they are lying, going sandy-coloured when lying on sand or black when lying on dark rock. There is a simple way of telling that colour changes of this kind work through the eye. It is only the upper surface of a flatfish that changes colour and that is where the eyes are situated. And, if the fish goes blind it cannot change colour. Many frogs and lizards can change colour according to the background. If they go blind they cannot change colour, but with frogs the moistness or dryness of the air around them also influences their colour.

It is hardly enough to say that it is the light falling on the eye that causes the change in an animal's colour. There must be some way of carrying the message to the skin where the chromatophores lie. This may be through the nerves or through hormones. If a flatfish damages one of its nerves running to the skin, the patch served by that nerve goes dark and stays dark while the rest of the skin changes colour. Changes of colour due to hormones are more common in invertebrates, like the Aesop prawn. Prawns belong to the Crustacea, a group which includes crabs, lobsters and shrimps. These have eyes on stalks. Their colour changes have been much studied and we know that they are caused by hormones given out in the eye stalks.

The word 'hormone' was coined, from a Greek word meaning to excite, in 1906. It was some time before it passed into common usage. A hormone is a fluid given out by a cell or a gland into the blood; it travels around the body and at certain places touches off, or excites, a particular process. That is why hormones are often called chemical messengers.

Underneath the brain of vertebrates is the pituitary gland, which gives out

several kinds of hormones. If the pituitary of a frog is removed by surgical operation, the frog cannot change colour. It goes black and stays black even on a white background.

Some insects – the stick insect is an example – can change colour to suit the background. The hormone responsible for the change is given out by its brain. Crustaceans have a gland lying just behind the eye in the eye stalk. This gives out a hormone that makes the chromatophores work. This has been proved quite simply. A shrimp that has lost its eye stalks by injury can no longer change colour when its background changes. Moreover, if we inject the blood from a shrimp living on a black background into a shrimp living on a pale background, it will go black. The obvious next step is to ask where the pigments come from, how they originate. The biochemistry of animal pigments has by no means been fully explored but the basic principles have been worked out. One group of pigments are known as the carotenoids, from their having first been noticed in carrots. They are responsible for many of the orange, red and brown tints. Only plants can synthesize them and animals obtain them from eating plants. Carotenoids are essential to any animal with eyes because the vitamin A in them is the source of the visual pigment in the retina. They are also responsible for the colours of the body in such insects as locusts, which obtain them direct from plants eaten, and in sea anemones and corals which eat animals that have fed on plants.

Another group of pigments are the ommochromes. These are waste products. They are responsible for colours in molluscs and in crustaceans and insects but are excreted by vertebrates. The pterins were first discovered in the wings of butterflies where they give rise to white, yellow, orange and red. Pterins are also found in the chromatophores of some crustaceans and are the source of vitamins, including folic acid and riboflavin, in plants. Guanine is also excretory. It is the origin of some kinds of iridescence in insects and of blue in fishes. In the latter, minute crystals of guanine produce a scattering of light which, against a background of melanophores, gives blue.

There is another aspect of animal colours which can only be dealt with in general terms but which is interesting to think about. This is, that as we go from the poles to the equator the amount of natural colour, both plant and animal, increases. One has only to think of the tropical plants with their large colourful blossoms and the wealth of coloured insects and brilliantly plumaged birds, like the birds of paradise, hummingbirds and the many kinds of parrots. This is not to say there are no colourful plants and animals in the temperate regions but there are fewer of them. The same is true for aquatic animals such as fishes, both in freshwaters and in the sea. If this were not so, aquarists would not be so ready to stock their aquaria with tropical fishes. So far as the sea is concerned the so-called sea gardens of coral reefs speak for themselves. The counterparts of the tropical marine animals that are to be found in temperate seas are to some extent colourful but they are neither so numerous nor their colours so intense. The sun, by shedding its rays and warming the earth, speeds up the metabolism, the chemistry of plant and animal bodies, and with it the production of pigments.

There are exceptions to this. Deserts for example, seem to favour the production of dull light colours, sandy to brown. This may be partly a direct result of a warm dry atmosphere. Certainly the reverse seems to be true, that animals living in dark humid forests tend to be darker than those of their own or nearly related species living outside the forests. On the other hand, both may be the result of natural selection, whereby anything other than a sandy coat in a desert, or a dark coat in a dark forest, leads to animals not coloured this way being killed off.

It is unwise to make categorical statements about animal colours or the effects of these colours. All too little has been directly tested. Some instances in which colour acts as camouflage have been tested, by experiment or close observation. In too many others our conclusions are based on theory, which may later be proved wrong. An example will help to make this clear.

Octopuses, cuttlefishes and squids have an ink sac and can squirt out ink. For about a century everyone said that these animals, when alarmed, squirted ink which acted as a smoke screen behind which they made their escape. The truth came out a few years ago when somebody put a captured squid in a tub of seawater. The squid shot out its ink which for some seconds hung like a

Squid Squid are related to octopuses and both are shell-less molluscs, distant relatives of snails, oysters and clams. Unlike these shelled relatives squid are active, powerful and fast swimmers. They also have well-developed eyes. One other outstanding feature of squid is the ability to change colour. This the squid does by using small pockets of pigment in the skin, each equipped with a ring of radial muscles. These are known as chromatophores and by the expansion and contraction of different colours in the chromatophores a squid can switch from one colour to another, or one pattern to another, in a fraction of a second.

An American scientist caught a 2-inch scarlet squid in a net. He straightway lifted the net to his face to take a close look. He thought the squid had gone until he noticed what looked like a hole in the net. It was the squid, now a pearly white, which is the usual escape reaction of squids. They contract all their chromatophores, so completely losing any vestige of colour. Their bodies then become translucent (semi-opaque) and appear colourless or pearly white. When squid are swimming in shoals a disturbance of the water, such as when a rock is thrown in, sends them all to the seabed where their colour changes to sandy. Lying absolutely still, they are practically invisible.

squid-shaped blob and then slowly dispersed throughout the water. Meanwhile the squid had apparently vanished. It was later found lying on the floor of the tub, colourless and motionless. It had shot out its ink, instantly lost all colour, and had dived to the bottom, leaving the ink blob looking in shape like the animal that had made it.

This is known as protective behaviour, which is something often observed in animals with protective colours, stripes or patterns. An example is seen in certain caterpillars that infest pine trees. When young they are longitudinally striped and lie parallel to the pine needles. After their last moult their bodies are cross-banded. Then they take to resting on the twig so that the bands blend into the bases of the needles.

Colours of animals are by no means as constant as we normally suppose. When we say the colour of this or that species is red or brown, it by no means follows that all members of the species will be that colour. The likelihood is that within that one species there will be as many variations in colour, or shades of colour, as in the hair of human beings.

There is a wild dog that ranges across the northern hemisphere variously known as the common wolf, timber wolf, grey wolf or red wolf. All represent one and the same species. Wolves in the Arctic are often white, on the North American plains usually grey, in Florida often black, and this does not exhaust the colours and shades. Wolves may also be buff, tawny or reddish. Even a grey wolf is not truly grey. Its coat is made up of white, black, grey and brown hairs. A wolf may be a different colour after a moult than it was before. In some districts the wolves may be mainly grey but there may be some that are black, red or white. In a single litter there may be pups of different colours.

The golden cat or Temminck's cat, of Southeast Asia, has colour phases. It may be red or grey, spotted or unspotted. All four may occur in one litter and each may change to another after a moult. The serval, an African wild cat, is yellowish with a bold pattern of black stripes and spots. Sometimes one animal has the stripes replaced by a fine powdering of specks. This used to be thought a distinct species and was called a servaline.

Colour may vary with age, with the season, with the sex, with diet, or it may be due to a straightforward colour change which may be ephemeral, transitory or semi-permanent. Although the examples given here of colour variation are all mammals, what is said can be applied to other groups of animals and with even greater effect lower in the animal scale.

Greater Kudu See page 104.

Elephant See pages 36 and 88.

African cardinal beetle Very few people notice cardinal beetles. The adult is usually a shade of red and for the most part it shelters under loose bark. The female lays her eggs in the wood of tree trunks. The larvae hatching from them are white maggots that feed on the wood for up to three years before changing into adult form.

An insect that is red, yellow or black, or has two or more of these colours in combination, usually has glands that give out something unpleasant, such as prussic acid. The cardinal beetle has not been tested for this but there is little doubt it contains either prussic acid or a chemical equally unpleasant. The burnet moth described on page 124 is another. When a young bird picks up one of these insects it soon drops it again. What is more, it shows by its actions that it has every reason to regret interfering with the insect. It rubs its beak on the nearest stone or twig as if trying to clean it, runs about and shows annoyance with any other bird it meets. It may be that the cardinal or the other insect concerned is injured, or even killed, in this encounter but it does mean that the bird will hesitate to touch a red insect in future.

16

Brave red colours

We live in a world that is mainly green and blue. The green comes from the grass and the foliage of bushes and trees. The blue is in the sky and, near a coast, in the sea. Against the green, especially, red shows up in a striking manner. Think also how the red of a sunset catches the eye and holds us transfixed, gazing at it until it begins to fade.

In many other ways, also, red has always been recognized as an outstanding colour. Soldiers today usually wear rather drab uniforms, except during some spectacular ceremony, when they don colourful uniforms, as they did years ago when battles were mainly hand-to-hand fights. Even when soldiers' uniforms were of other colours they often had red facings, or facings of some other bright colour. The pattern of these indicated at once what regiment the men belonged to. They were, in a sense, signals carrying an immediately intelligible message.

In fact, an outstanding colour can almost always be said to carry an 'instant message', that is, it tells us something important at a glance, such as the international stop and danger signal. This is just as true among animals as among humans. We find so often that an animal coloured bright red either carries a poison or is in some other way unpleasant. Such animals, bearing their own danger or warning signals, seem to be fully aware of it. They do not trouble to hide but seem instead almost to flaunt their scarlet colours, as if saying 'Touch me if you dare'. Red is the colour used by men to portray importance – look for example at the robes of a Cardinal or royal robes.

Among the animals, especially among birds, it is nearly always the males who wear the bright colours. This, it is usually said, is to attract the females, but modern science has cast a doubt on this view. It may be true, though more often and more obviously the male animal uses bright colours to warn other males not to trespass on his territory. Reds, more than most colours, are then used.

Naturally, the use of colours for this or any other purpose will be ineffective if the species does not have colour vision. It used to be said that all furred animals, other than monkeys, apes and man, lacked colour vision. In the last twenty to thirty years, however, it has been shown that many of them, such as horses and dogs, have at least partial colour vision. Nevertheless, we can in general say that species that wear dull colours either have little colour vision or none at all, and that those with bright colours have good colour vision.

Finally, although this chapter is devoted more particularly to animals coloured red, it should be remembered that there is more than one shade of red, from brilliant scarlet through orange to pink, with many other intermediate shades such as reddish brown. In one field of behaviour more particularly, namely blushing, we can see the transition from pink to crimson or scarlet. Two examples are included in this chapter, the firemouth and the hooded vulture, which illustrate an important principle: only exposed skin blushes. In birds and mammals, with the body covered by feathers and hair respectively, blushing can occur only where there are areas of naked skin; in man it is confined to the face and neck.

Common cardinal Quite a number of American birds have been called cardinals but most of them have red only on the head and neck. The reddest of all is a bird of the finch family, 8 ins long, known as the common cardinal, of the United States. Even its heavy beak is red, although it also has some black feathers.

Only the male cardinal is so completely red. The female is mainly brownish and only her beak is red. Where the male and female of a species of bird are very different in colour it is usual to say that the male's colours attract and please the female. This is hardly necessary here, for a pair of cardinals will stay together the year round. More probably, the striking red colour of the male is used mainly against other males.

Male cardinals are apt to bicker among themselves and they use their red feathers in displaying aggressively at each other. This aggressive display is especially marked at the start of the breeding season, when the birds are marking out their territories. The display is also used in courtship. A male will perch on a twig above a female, stretch out his neck, raise his red crest and sway from side to side or bow to her as he sidles towards her.

Scarlet ibis An ibis is a wading bird with a long neck and a long, slender, downward-curving bill. Ibises live near water in tropical and subtropical countries, feeding mainly on small aquatic animals. There are several different species, the most famous being the sacred ibis (see page 96) which was venerated by the ancient Egyptians. Some ibises are white, others have a dark plumage with a metallic sheen.

The most beautiful is the scarlet ibis, about two feet long, a South American species ranging from Venezuela to Brazil. Sometimes it strays north to the West Indies or, more rarely, to the southern shores of the United States. It nests in colonies in trees, which then seem to be spattered with blood.

The local aboriginal peoples used to kill it for its flesh, although this is rank and oily. It has also been slaughtered mercilessly for its feathers, so that the large colonies that used to be found in the mangrove swamps have been much reduced in numbers. Fortunately, a sanctuary has been created in Trinidad and there the birds are becoming more numerous.

The scarlet ibis feeds in swamps, in muddy estuaries and on mudflats, eating fish, insects, molluscs and crustaceans, many of these being picked out of the water or out of mud with the long bill. When in captivity it tends to lose its brilliant scarlet, which seems to be a colour derived from its food.

Blood-red butterfly Red is not a common colour in butterflies and usually when it is present it occurs in patches and spots. This blood-red butterfly, *Cymothoë sangaris*, 2 ins across the spread wings, of the forests of Central Africa, is an exception. Its near relatives are also red but they are orange red, not the startling red of this one species which, incidentally, has no common name. Moreover, it is only the male *Cymothoë* that has this bright colour. The female is a duller red or reddish brown marked with dark wavy lines.

Where male and female are coloured differently, as is so often seen in birds and fishes, as well as in insects, it is convenient to speak of a sexual dichromatism. The basic principle behind this is that the female, being the bearer of the next generation, must receive more protection than the male to ensure the survival of the species. So she is more effectively camouflaged. Colours are needed to bring the sexes together, either because male and female occupy different parts of the habitat or for other reasons. It is more economical, from the point of view of the species, that the males should wear the striking colours because they are the more expendable.

Red mite Since Anglo-Saxon times the term
'mite' has been applied to species of very
small creatures related to the spiders. The
best known of them is the cheese mite. Later
in the Netherlands, the same name was given
to a very small Flemish copper coin, perhaps
because of its size, and we meet it in this sense
in the Bible, in the story of the widow's mite.

Because most mites are invisible or only
just visible to the naked eye, people know
little about them. Yet they are everywhere on
land and in fresh water. Insects are numerous
enough but mites, which are arachnids, are
probably just as plentiful. Astronomical
numbers live in soil and are important for
keeping the land fertile. They feed on minute
fragments of plants and so break them down
even further to enrich the soil.

Some mites are red. In eastern and
southern Africa there are what may be called
giant mites, nearly half an inch long. Their
rounded bodies and four pairs of legs look
velvety, and because of their size they are
sometimes called red spiders. One of these
red mites, of East Africa, is shown in our
picture on a flowering desert plant. Their red
colour seems to have no biological value.

Galah The galah, also known as the pink-
breasted cockatoo, is one of several beautiful
Australian species. A flock of galahs, several
hundred strong, makes a wonderful sight
wheeling through the air, the sun alternately
catching the birds' silvery-grey backs and
rose-pink breasts. The galah, about 14 ins
long, feeds on grass seeds, plants and bulbs
dug up with the beak. With the spread of
farming in Australia the galah population has
become very numerous, and the birds are
often seen in parks of the large cities.

Other spectacular cockatoos which form
large flocks include the sulphur-crested
cockatoo (see page 82), the little corellas and
the cockatiels.

Lesser flamingo The lesser flamingo of Africa
and India lives in immense flocks. During the
breeding season the colonies of lesser
flamingos that assemble on the alkaline lakes
of the Rift Valley in East Africa, number
hundreds of thousands, sometimes more than
a million – a truly amazing sight! Although
there is some white in the plumage, the
feathers are mainly pink or deeper red, the
legs also being pink and the beak carmine red.
The pink colour is more pronounced than in
the somewhat larger species with which it is
often seen – the greater flamingo.

When feeding, the lesser flamingo lowers its
head to the water surface and sweeps its bill
from side to side, scooping up small
crustaceans and sieving out the mud. The
bird's pink colour is due to a carotenoid or red
pigment contained in the microscopic water
plants eaten by the tiny crustaceans on which
the flamingo feeds. It will eat other types of
food as well; and in zoos, where the diet is
necessarily more varied, the distinctive pink
colour tends to fade.

Red Bishop (left) On the grassy plains, or savannahs of Africa live a number of different kinds of weaver birds. These are related to sparrows but are more colourful. Some of them have short tails, possess a good deal of red in their plumage and are called bishops. The others are called widow birds or whydahs and are mainly black. In the breeding season the male widow birds grow very long tails.

Outside the breeding season the bishops live in large flocks flying about the savannah, wherever the grass seeds are ripening. Then male and female look much alike. As the breeding season approaches the males grow bright red feathers and the flocks split up. Each male takes over a territory, courts a female and patrols the boundary of his territory, flying slowly, vibrating his wings noisily and rapidly. Should another male draw near he lands on a perch, points his bill upwards and fluffs out the red feathers on the nape of his neck, on the collar or ruff, and at the base of his tail. This is a sign to the other male that he must keep out.

Frigate bird Frigate birds, also known as 'man-o'-war birds', spend almost all their lives flying over the sea. They seldom stray far from their breeding grounds on oceanic islands, different species being found in the Atlantic, Pacific and Indian Oceans. In flight the long black wings, up to 7 ft across, the long

hooked bill and forked tail provide an unmistakable silhouette.

Speedy and very manoeuvrable in the air, frigate birds swoop on flying fishes or snatch food from the surface. They also harass other sea birds such as gannets and boobies, forcing them to disgorge their prey and catching it in mid-air.

The male frigate bird is smaller than the female and differs from her, too, in possessing a bright red throat patch which is actually an air sac. At the beginning of the breeding season he inflates this until it takes on the form of a large balloon covering most of the breast. With wings spread and sac fully inflated, the male perches on a branch and postures to the female as she flies overhead.

Attracted by the scarlet throat sac she flies towards the male who claps his bill and rattles his quills, his body quivering all the while in his excitement – a mood soon communicated to his mate. So the scarlet sac can be said to stimulate the female to breed.

African grey parrot The grey parrot is one of the most popular pet birds and has been so since at least the sixteenth century. King Henry VIII owned one, and the skin of a pet grey parrot kept by the Duchess of Lennox and Richmond, a hundred years later, is believed to be the oldest stuffed bird in the world.

Less colourful than many other members of the parrot family, the African grey is still attractive. It is mainly pearl grey with a white face, black beak and a red tail. But its popularity comes chiefly from the bird's ability to mimic sounds and sometimes actions. A good talking African grey will have countless words, phrases, snatches of songs, whistles, and all kinds of imitations of mechanical sounds in its repertoire.

Some African greys used to be known as king grey parrots; these had some pink feathers on the body as well as in the tail. 'Kings' were much sought after because they were believed to be much better talkers than the rest. The truth is, however, that these pink feathers come either with old age or because of a deficiency in the diet.

Crimson rosella There are 316 parrots, cockatoos and other members of the parrot family in the world and a fifth of them live in Australia. In most of the Australian species green and red predominate. The red occurs mainly in patches but in the rosellas, more especially, red is predominant. The crimson rosella, shown here, is one of the most popular parrots of Australia. Its bright crimson and blue are so startling that there could be little question of the bird being camouflaged. Since it lives in the damp woodlands of the eastern parts of Australia, where the dense foliage hides it anyway, camouflage is hardly necessary. Moreover, because the rosella nests in hollows in trees, it is most completely hidden at the time it is most vulnerable, when sitting on its nest or attending to the young. On the other hand, it is probably true to say that because there are so many highly coloured blossoms on the trees, the striking colours of Australian parrots tend to tone in with these. This in itself would be protective. Some of the colours of parrots must also enable parrots to recognize members of their own species, especially where they live in mixed flocks.

Cock-of-the-rock The brilliant orange plumage of the male common cock-of-the-rock, a native of South America, is used on social occasions. A group of males will gather on a display ground, each bird clearing a small area of any twigs and leaves that might impede movement. Then all the assembled males strike postures which they hold motionless for several hours. At this stage of the proceedings, there is none of the dancing, bowing and bobbing that most male birds of other species adopt when courting. Whether these statuesque poses impress the plainer brown females is not known.

The male of a related species, the Peruvian cock-of-the-rock, possesses bright red plumage, and whereas the head crest of the common cock-of-the-rock leaves part of the bill exposed, that of the Peruvian species curves round so as to cover the bill almost entirely. Both species belong to a family that also includes the colourful cotingas, which are somewhat smaller, and the rather larger umbrella bird.

Ross's turaco All over Africa are found large fruit-eating birds with long tails and short rounded wings. They are brilliantly coloured. Ross's turaco, of tropical Africa, shown here, is a glossy blue, with red feathers in the wings and a red crest. The red crest and red patches on the wings are used in courtship. The crest is raised, the birds bow to each other, flick their wings and flirt the tail.

The main reason for the turacos' fame is, however, an unusual belief about their red feathers. For more than a century it has been said that the red pigment in the feathers is washed out when it rains. This is entirely untrue.

It is, perhaps, because of this strange and mistaken belief that a good deal of chemical investigation has been devoted to the pigment in the red feathers. This has shown it to be a copper complex which forms a pigment found nowhere else in the animal kingdom.

Fighting fish Many fishes fight, usually over territory. They swim around each other with their fins erected to the full, trying to show their coloured patches to best advantage. Then they may grip one another with the jaws or bite off pieces of fin. Of all the bellicose fishes in the world the most famous is the fighting fish of Thailand.

A wild fighting fish is about 2 ins long, yellowish brown with indistinct dark stripes along its flanks. In the breeding season the males become darker and the rows of metallic green scales in the flanks are brighter. They also have red on the fins. But even they are dull beside the fighting fishes seen in aquaria.

The Thais have been breeding these fishes for colour and fighting qualities for centuries. Owners arrange fights, bets are placed on the outcome, and the fishes then fight tenaciously. In the wild, fighting fishes will do battle for as much as fifteen minutes. Those specially bred for their fighting qualities will fight for much longer, seldom giving up in under an hour and usually going on for up to six hours.

With the increase in the fighting abilities of the domesticated varieties have come changes in their form and colour. The fins have become flowing and veil like and the body may be coloured blue, lavender, green, red or purplish blue, usually with red fins.

Scarlet coral fish Coral reefs are veritable sea gardens, highly colourful, and the fishes that inhabit them are equally colourful. The fish shown here is a scarlet coral fish and lives, appropriately, in the Red Sea. They belong to a group known generally as anemonefishes because some of their members live in close association with large sea anemones. The corals seen here are gorgonians or horny corals.
Instead of the stony skeleton of the true reef corals, the branches of a gorgonian have a stiff, horny axis.

Cardinal tetra (Top left) Tetras are very small fishes living in forest pools of north-eastern South America. Most of them are only an inch or so long, are very colourful and the colours have been incorporated into their common names. So we have the red tet or flame fish, the yellow tet, the dawn tetra (which is pink like the sky at dawn), as well as the neon tetra that has such bright colours it looks like a tiny, animated neon sign. Among them is the cardinal tetra. It is closely related to the neon tetras and indeed has a blue 'neon' band which, from some angles, is more prominent than the red that gives it its common name. The red is, however, more

widespread, covering the lower flanks and the whole underside of the body; and the colours are the same in both male and female. Scientists have not yet discovered the real purpose of the colours of tetras.

Spurge hawk moth (Top right) Hawk moths are large, the largest having a wing span of several inches. They also fly rapidly, usually at night, and most people do not see them unless they come in through the open window to fly around the light in the room. Only a few species are active by day.

The spurge hawk moth shown here is

nocturnal, resting by day. Should it then be disturbed, it spreads its wings and goes in for what is called a defensive display. Whereas most moths quiver their wings when disturbed, the spurge hawk moth dances or jigs about but keeps its wings spread and rigid. In this way it flashes the bright red patch on each of its hindwings.

So long as the moth is relaxed and at rest, the red on the hindwings or underwings remains hidden. When a bird settles near the moth and is likely to make a stab at it with the beak, the moth suddenly opens its wings. This is startling enough to the bird, but the sudden appearance of coloured patches

increases the surprise effect. The fact that they are red, a warning colour, may throw the bird completely off balance and it will fly away, leaving the moth unmolested.

Sea hares (Above left) These are small, sluglike marine molluscs that browse the seaweeds among which they live. Usually they are not much more than 2–3 ins long, of a brownish colour. On their head is a pair of tentacles. In addition there is a second pair of tentacles, carrying many sense organs, that are rolled inwards. They give these molluscs something of the appearance of crouching hares. Sea hares are worldwide, occurring especially in tropical and subtropical seas, and all look very much alike. The largest live off the coast of California. All lay eggs in tangled pink or orange ribbons. Observation of one Californian giant sea hare, weighing 5 lb 12 oz, showed that it laid 470 million eggs in six weeks.

African grasshopper Rules are made to be broken, it is said, and this we find only too often in nature. Yellow, orange or red with black bands are warning colours and these are well shown in the cinnabar moth caterpillar. The lubber grasshopper of East Africa is slow and clumsy in flight, and it makes no attempt to hide, almost as if courting disaster. It is, however, protected because it gives out a caustic yellow fluid from the joints of its legs and thorax. It draws attention to this by flashing its orange underwings (the hind pair of wings). The abdomen, moreover, has black bands, so here the bands and the warning colour are side by side instead of alternating, as in a wasp or a cinnabar caterpillar.

Firemouth cichlid There are many species of freshwater fishes belonging to the family Cichlidae. One that lives in Yucatan and Guatemala, in South America, has a red belly. The colour is more correctly described as a fiery orange and it extends from the base of the tail to the mouth, or even into the mouth, hence its popular name.

Something similar to the human habit of blushing with anger and paling with fear has happened with the two firemouths in this picture. Both are males and the one on the left is the owner of a territory. The firemouth on the right has wandered across the boundary of the territory and the owner is blushing a brighter red on his undersurface. The trespasser, not wishing to stand and fight, has gone pale. This is a signal that he is going to retreat and hide himself. Human blushing is caused by an increased amount of blood flowing into the face, into the fine blood vessels under the skin. Fishes that 'blush' change their colour by using pigment cells in the skin.

Reed warbler Small birds that begin life as nestlings, unable to leave the nest for some days, usually hatch blind and almost naked. The four nestlings of the reed warbler, of Europe, shown here in their cup-shaped nest among the reeds, are only a few days old. Their eyes are still unopened. The body is clothed only in sparse down. The stubs of the first feathers have broken through the skin which is pink, a colour due to the tiny blood vessels beneath the skin.

There are some 400 species of warbler in the Old World, more easily differentiated by their song than by their appearance. A dozen of these are reed warblers. All reed warblers live in marshy areas and build their nests on reed stems above ground or water level.

Hooded vulture The hooded vulture is found over most of Africa. It is small by vulture standards, brown except for the head and neck which are naked, except for a sparse layer of down. And it blushes.

Like other vultures, it feeds on carrion but also eats insect grubs, grasshoppers and locusts. It is quick to find an animal carcass and is usually first on the scene when an animal dies. It must, however, wait for one of the larger vultures to arrive to rip the carcass open with its more powerful beak.

Even then it is not easy going. Vultures are inclined to squabble over the more tasty bits of a carcass. So a small vulture is pushed aside by the larger birds and has to take what it can while the others are not looking. When dealing with a bird nearer its own size, such as the Egyptian vulture, the hooded vulture starts to assert itself. The pink skin on its face and neck becomes a deeper red as it tries to hustle the other bird away, as if it were blushing with anger.

Alpine newt (Left) The alpine newt lives in mountainous districts of central and southern Europe. The male is just over 3 ins long, the female nearly half as long again. This newt is found in great numbers in stagnant or running water, sometimes even in a puddle. There is little to distinguish it from many other kinds of newts in the northern hemisphere, except its colours. The back is a marbled grey to black, the flanks bluish and the underside orange.

Starfish and scallop The animal kingdom can be divided into the vertebrates – animals with backbones – and the invertebrates – animals without backbones. We can think of the invertebrates as being of two kinds, the lower and the higher. The lower invertebrates live mainly in the sea and either have no eyes or very simple ones that can do little more than distinguish night from day, darkness from light.

Clearly, therefore, they are unlikely to use colour signals to communicate with each other, which means that their colours do not have the same practical values as in animals that possess eyes. Yet some of the most colourful of all animals are found among the lower invertebrates living on the seabed where, because of the absence of bright sunlight and the general murkiness, everything around looks a dull greyish green – at least to the eyes of the underwater swimmer. In the picture right is a scarlet starfish which eats the flame scallop, a bivalve shellfish seen to the right of the starfish. The red in both these marine animals is due to the pigments known as carotenoids. These are manufactured by many plants and the animals that feed on the plants take up the colour. So also do the predators that in turn feed on these animals. Flamingoes are pink, but only if they get their proper food containing carotenoids.

Slate-pencil urchin The hedgehog, with its rounded body covered with spines, used sometimes to be called an urchin. When a small animal with rounded body covered with spines was found living in the sea it was, quite naturally, called a sea urchin. The sea urchin is related to starfishes or sea stars and its spines are usually sharply pointed. There are, however, some species in tropical seas in which the spines are blunt and thick, and triangular in cross section. Indeed, their spines look very like the old fashioned and now obsolete slate pencil. The fact that there are few people today who can remember what a slate pencil looks like makes the common name less helpful than formerly; but urchin was the original name.

bull-mouth helmet shell, coloured bright red, which lives in the Indian Ocean. At the front it has a siphon to draw in water for breathing, and there are two tentacles, each with an eye at the base, protruding from the head.

Ostracods (Left) Ostracods are crustaceans with a bivalved shell that encloses the body, trunk and limbs. The shell is hinged at the top. Two pairs of antennae projecting at the front serve both as feelers and also as limbs for swimming. Ostracods are found in all seas and in rivers and lakes. Some are planktonic and strain small food particles from the water, others are scavengers on the bottom and the larger of them may feed on small shrimps and other small animals. Most ostracods are less than 3 mm long. The largest known is the size of a cherry and the one shown here is 6 mm long. It lives in the tropical Atlantic at depths of 300 to 500 fathoms.

Purple sea fan Many marine animals are plant-like, living fixed to one spot once the larva has settled. For them, colour, so far as we can see, is unimportant. It can hardly be camouflage since sea fans tend to stand out not only by reason of their shape but also because of their colour. Sea fans are soft corals related to reef corals, sea anemones and jellyfishes. The colonies of polyps, of which they are formed, are armed with numerous stinging cells. Possibly, therefore, their bright colours represent a warning coloration. The purple sea fan shown right lives in Indonesian seas.

Bull-mouth helmet shell There are two major groups of invertebrate animals known as molluscs which differ markedly – the bivalves and the gastropods. Bivalves, characterized by a shell divided into two halves, include oysters and clams. They move about little or stay in one place, feeding mainly on microscopic plants and animals. The gastropods, commonly known as snails, and including also whelks, limpets, cowries and conchs, have a coiled shell and are much more mobile, crawling on a fleshy foot; and they are often carnivores, feeding especially on bivalves. One of the latter group is the

African elephant There are two species of elephants living today: one in Asia, the Asiatic or Indian elephant, and one in Africa, the African elephant. The second of these is slightly the larger and may reach 6 tons. The most obvious difference between them is in the ears, which are much larger in the African elephant. Other differences are that the Indian species has a domed forehead as against the sloping forehead of its African relative. There is also a difference in the trunk. The African elephant has two 'fingers' at the tip of the trunk, against one in the Indian elephant.

By no stretch of imagination could the elephant be called colourful. The Indian elephant is usually described as being dark grey to brown The African elephant is usually said to be greyish brown. So there is not a lot of difference there. Moreover, both species have the habit of wallowing in mud or dusting themselves. As a result, both in Asia and in Africa, elephants tend to take on the same colour as the earth in the region in which they live. Often this is reddish. The picture shows an African elephant in the Tsavo Park, in Kenya, plastered with the red mud in which it has wallowed.

Orang-utan This auburn-haired baby orang-utan will grow into an inoffensive giant weighing anything up to 220 lb, depending on whether it is a male or a female and how long it lives. The orang-utan, whose name is derived from a Malay word meaning 'man of the woods', is one of the anthropoid or man-like apes, only exceeded in size by the gorilla. But whereas the gorilla, an inhabitant of Africa, spends most of its time on the ground, the orang-utan is a tree-dwelling ape, living only in dense rain forests on the islands of Borneo and Sumatra. It sleeps in the trees and swings from branch to branch with arms that are much longer than its legs.

The orang-utan's skin is greyish, the face is naked, and the body is thinly covered with reddish hair. But neither its coloration nor its solitary life habits have prevented this sad-looking ape from being mercilessly hunted in the wild. Females are frequently killed and the babies captured alive for transportation to zoos overseas: and orang-utans in captivity, deprived of all opportunity for exercise, grow heavy and lethargic, often dying of infectious diseases. Not surprisingly, the orang-utan is one of the world's threatened species.

Water vole There are some remarkable colours that are not found on the outsides of animals. The inside of the mouth and throat of nestling birds are often beautifully coloured. The green turtle is so named for its green fat. The marine garfishes or needlefishes, have bones that are green, and so is their flesh at times. We expect teeth to be white but, surprisingly, some animals are exceptional even in this respect. There are, for example, many red-toothed shrews and some of the larger rodents have orange-coloured teeth, as in the water vole of Europe.

Red fox The familiar fox of Great Britain is also found across Europe and most of Asia, as well as North Africa. For as far back as one can trace, it has been called simply the fox. In the seventeenth century people went out from England to settle in North America, where they found an animal looking exactly like the fox they knew in the homeland. In time they discovered another North American fox, the grey fox. So the other one became known as the red fox.

Today, it is fast becoming the fashion in Britain to speak not of 'the fox' but of the red fox. This is because so much more is now known about other foxes in other parts of the world that we need to be more specific.

There is every justification for calling both these animals 'red', because the coat of the average fox is a distinct red. There is, however, so much variation, especially in the coat of the American red fox, that it gives us a good example of how a colour can vary within a single species.

In the American red fox the coat may be silver or black, and there are other patterns too. So the red fox may also be known as a silver fox or a black fox. In Eurasia the coat may be a distinct red or it may be brown or sandy, or even a blackish red. Whatever the colour of its coat the fox thrives equally well, for in no case does it serve as effective camouflage.

Here, then, is a good example of the fact that while some colours in animals may have a definite function, either serving as camouflage, or to give signals, there are some that seem to have no biological value.

Superb starling The superb starling of East Africa, one of the so-called glossy starlings, seems almost determined to show off its fine feathers. It lives in small flocks in the acacia grasslands, is sociable and seemingly fearless of human beings. Like other starlings it feeds on insects and berries, but it has a particular preference for termites. That is one reason why it leaves its roost early. The termites forage at night and return underground before the sun is up. Warblings at the roost half an hour before dawn indicate that superb starlings are already awake and at first light they fly down from the trees to catch the home-going termites.

Although far more spectacular than common starlings, these African birds display the gregarious habits that are characteristic of all members of this widespread family; and the metallic blues and purples of the plumage glint when the sun catches them.

Brilliant blue decoration

The blue of the sky is caused by the scattering of sunlight by particles in the upper atmosphere. The ocean is divided into two zones. One is called the neritic province and includes the shallower waters fringing the coasts where the bottom lies from 0–450 ft deep. The second province is the ocean proper, beyond the neritic province. Neritic waters look green because particles of silt brought down by rivers are still held in suspension and with the silt is much plant plankton. The green is due to a combination of scattered blue light and yellow pigments in the particles and plankton. The water of the ocean province beyond is clear and looks blue because there the sunlight is scattered by the much smaller particles of the molecules of water.

Although there is much blue in parts of our environment, the sky and the oceans, blue is a fairly rare colour in animals. For example, there are about 3,000 species of sponges and only one is pure blue. Even that is caused by bacteria living in the tissues of the sponge.

From what has already been said it follows that blue animals are found mainly in the sea and especially in the open ocean. Many jellyfishes are blue, such as the powerfully stinging jellyfish known as the Portuguese man-o'-war, which is often called bluebottle. Some of the larger oceanic fishes have blue backs, and the scourge of the western Atlantic that massacres huge shoals of fishes annually is the bluefish. Blue is, however, most notable in small animals forming the plankton in the top layer of the ocean, the upper few inches.

In land animals as well as freshwater animals, blue is not common. A few butterflies are wholly or almost wholly blue, but the colour occurs more usually in spots, small patches and stripes, as if purely for decoration. It may range from very pale powder blue to the almost black indigo blue. One group of birds has markedly more blue than most, namely the kingfishers. By a paradox, some of the best examples of blue birds are jays belonging to the crow family, notable for so many black members.

The blue of kingfishers, jays and parrots and in some monkeys, such as the mandrill, is structural and produced by a scattering layer backed by a black background. This is known as Tyndall blue. Blue eyes are produced by a scattering layer of protein particles in the iris backed by a layer of melanin. Brown eyes result when the blue is masked by melanin granules among the layer of protein particles. There is one very striking example of a bird that strongly prefers blue to other colours. This is the satin bowerbird of eastern Australia. The male builds a bower on the ground consisting of a platform of sticks, in which he plants a double row of sticks. Within this avenue he displays to the female. The young male has a green plumage which later turns to a dark blue. His eyes are a light blue. The female has a green back and a brownish spotted breast.

The male decorates his bower with all manner of flowers, snail shells and other objects such as pieces of glass and string, paper and feathers. His preference is for blue: a blue matchbox or blue envelope, for example. He also daubs the sticks of his bower with the blue juices of berries. One male satin bowerbird kept in an aviary was found to be killing blue finches, kept in the same aviary, and using their carcases to decorate his bower.

Blue jay (Below) The blue jay is a colourful bird whose range extends throughout the eastern part of the United States and across much of southern Canada. Like other jays, it is a member of the crow family, a surprising fact considering that the better known members of the family – common crow, raven and jackdaw – are almost entirely black, although in strong sunlight the feathers are seen to be shot with blue and purple. In the jays the black is confined to patches, bands or streaks, and the remainder of the plumage is highly colourful. The blue jay, a woodland species which is frequently seen in city parks and gardens, is bright blue above and greyish below, with a black collar and specks of black and white on wings and tail. Another colourful North American species is Steller's jay, which is blue all over.

Common jay (Bottom) Song birds sometimes have a strange routine termed anting. They pick up ants in their bill and rub the insects on the underside of their wings. It has been suggested that they are using the formic acid given out by the ants to drive vermin from their feathers, but this seems unlikely, and scientists are still not sure why the birds act in this way. Some maintain that a bird 'ants' to tone its feathers, but others have put forward alternative theories. Whatever the reason, one thing is certain: the common European jay, when anting, looks more colourful than at any other time, largely because it shows to the full the chequered blue and black feathers on its wings. The common jay builds its bulky nest in trees, usually oaks (acorns being favourite food), and the male shares in the incubation of the eggs.

Red-cheeked cordon bleu The name 'cordon bleu' is more familiar today in relation to cookery but it was first applied to a small bird. There are two species of such birds, similar and related to each other, and both live in Africa. One is the red-cheeked (or red-eared) blue waxbill, the other is the blue-breasted waxbill. The first of these is imported in large numbers for those who keep aviaries. Its native home is in tropical Africa. Its red

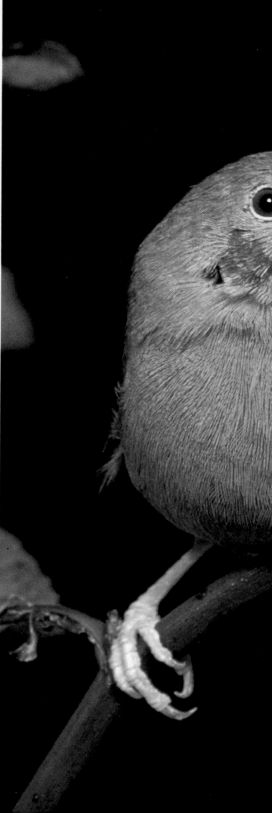

cheek is clearly seen in the left-hand bird in our picture. The other species lives farther south. Both feed on grass seeds and insects. The most interesting feature of the red-cheeked blue waxbill is the display of the male in courtship. He picks up a grass stem or a feather and flies up to where the female is perched and there bobs and twists, with feathers fluffed, towards his mate.

Malachite kingfisher Kingfishers comprise one of the most colourful families of birds, and the malachite kingfisher is a particularly spectacular species. Malachite is an anhydrous carbonate of copper. It is green but there is a blue form of it known as

azurite. The malachite kingfisher is only $5\frac{1}{2}$ ins long of which $1\frac{1}{2}$ ins are taken up by the dagger-like bill. It is found over almost the whole of Africa south of the Sahara, by quiet streams, rivers and ponds. It is, however, less easy to see, despite its brilliant colouring, than most small kingfishers. Its bill and feet are red, its back royal blue, its front pale cinnamon; and the malachite crest is erected when the bird is alarmed.

Kingfishers employ a variety of fishing techniques, some hovering over the water, others standing guard on a branch and then diving for their prey. The malachite kingfisher flits to and fro, skimming the surface and submerging only for a split second as it impales a fish with its beak.

Rainbow bee-eater Australia boasts many beautiful birds and the rainbow bee-eater ranges widely over the mainland and to the islands lying to the north, from Celebes to the Solomons. As its name suggests, the plumage contains a variety of colours but blue is more clearly visible, especially during flight, than other hues. It is seen here returning to its nest with insect food for the young. The nest, similar to that of the related kingfishers, is a simple chamber at the end of a 6 ft tunnel burrowed in sandy ground. There is no vegetational nest lining and the interior soon becomes cluttered with droppings and the remains of bees and wasps. Although the young are born on the ground they later join their parents in the trees where most of their food is to be found. Other species have similar habits, either catching their prey among foliage or on the wing. More than half of the two dozen known species live in Africa.

Common peafowl Birds do not like being stared at. It is, for example, easier to approach a wild bird for a closer look by averting one's eyes. By a strange contradiction, the peacock of Southeast Asia uses a multitude of 'eyes' in its efforts to attract and impress the peahen. Perhaps the large number of eye spots exposed when the train is spread removes any feeling by the peahen that she is being focussed upon.

There are two species of peafowl, the blue and the green. The blue peafowl, of India, is the best known because it has been taken to so many parts of the world, as ornamental fowl. They would be kept more than they are but for the male's loud wailing call. The peahen is a drab brown. The male's great train, made up of 150 feathers and several times longer than the body, is carried horizontally and erected in a fan when the bird is excited. The train feathers may take several months to grow following the annual moult.

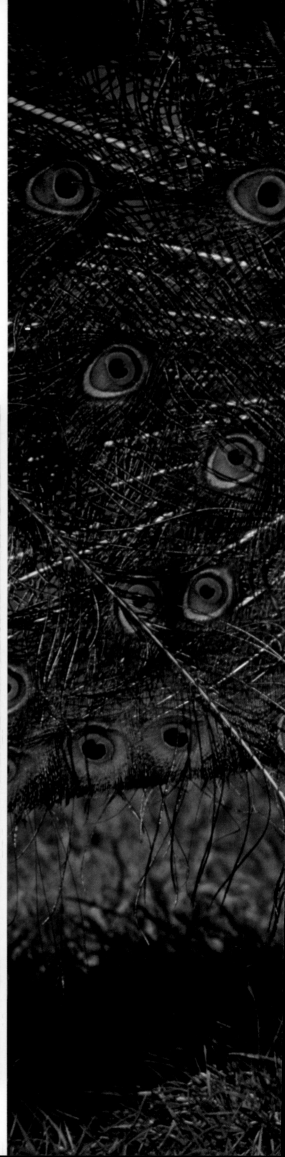

42

Flower mantis (Far left) These insects, known as praying mantises, sit erect on a flower or leaf holding the front pair of legs together, as if in prayer. These two front legs are crooked like a pair of half-closed jack-knives. As another insect approaches, a mantis extends these two legs and, like lightning, jack-knifes them around the insect. Then it eats it. Usually mantises are coloured and shaped to look like leaves or flowers. The flower mantis of Africa, shown here, goes one better. Its body is covered with knobs and as it sits in the middle of a daisy-like flower it seems a natural part of the centre. When young it can also change colour to match the particular flower on which it is perching. If disturbed, the adult flower mantis, which has wings, flashes these sideways and the spots on the forewings look like eyes. This helps to scare away a bird that has penetrated the mantis's disguise and is about to eat it, a result similar to the one produced by the 'eyes' of the peacock butterfly and the eyed hawk moth (page 48).

American butterfly (Above) The *Heliconius* butterflies of northern South America and Central America are noted for giving off an acrid odour. This is detectable by the human nostrils at a distance of ten yards. These butterflies, 2–4 ins across the wings, also have the habit of resting at night in one particular tree or shrub, and they return to it night after night. They include some of the most beautiful American butterflies. Some are black with bold red, yellow or white markings. Most are blue or green. The one shown here is a good example of a physical colour. The iridescent blue changes position on its wings as the butterfly is viewed from different angles.

Morpho (Left) No butterflies show such an expanse of iridescent blue as those of tropical America known as the morphos. As usual with butterflies, it is the males that are colourful and even they are cryptically coloured on the undersides of the wings; that is, the undersides are shades of brown and other dull colours in a broken and intricate pattern that resembles bark and similar backgrounds. As a morpho flies through a clearing in the forest, the sunlight catching its wings is reflected in blue flashes. Then, as the insect turns presenting its cryptic underside to an observer, it seems to vanish, only to reappear a split second later in a most baffling manner. This is known as the dazzle effect. It happens again as the morpho alights, when it raises its wings and brings them together over the back, so masking the bright colour on the upper surfaces of the wings. Any bird pursuing a morpho will be thoroughly confused. Most butterflies use this dazzle effect. The more brightly coloured they are on the upper surface the more they continue to use it so long as they are perched, but not actually resting, by repeatedly opening and closing the wings.

Common blue The inequality of the sexes is very marked in a small butterfly, only an inch across the spread wings, that is common in meadows and on grassy hills in Britain. The females are mainly brown and so look unlike the males in which the upper surface of the wings is some shade of blue. The larvae feed on a low plant of the pea family, the birdsfoot trefoil.

Most of the colours seen in butterflies are due to pigments. The blues and metallic shades are mainly structural. A butterfly's wings are covered with minute scales arranged along the veins of the wing and overlapping like tiles on a roof. They can be rubbed off as a fine powder leaving the wing transparent. The iridescence so often seen in the blue patches on a butterfly's wing is due to each scale being made up of thin films separated by material of a slightly different, refractive index. Sometimes it is due to minute ridges that diffract the light falling on the scales.

Mazarine blue (Top) This European butterfly belongs to a family that is worldwide and includes butterflies known as blues, coppers and hairstreaks. These names are alone enough to suggest a diversity of colour from one species to another. Members of the family are found from the tropics to the Arctic, and high up in the mountains beyond the timber line. Although all are small and many are minute, it is hardly possible to go out into the countryside when the sun is shining without encountering them.

Mazarine blue is a deep rich blue. The name came into use in the seventeenth century and may be associated either with Cardinal Mazarin of France or the Duchesse de Mazarin, probably the latter.

Plaice (Left) The flatfish known as the plaice is not one of the more colourful residents of the sea but it is a master of the art of camouflage. Lying on the sandy seabed, its pearly-white underside is lost to view and its upper surface takes on the colour of its surroundings. As it settles on the bottom, the plaice flicks sand onto its back and this provides even more effective concealment. All that is left to betray its presence are the two blue eyes which lie on the top of its head and these can be turned independently in different directions. As a larva, the plaice looks much like any other fish, with an eye on either side of the head, but in due course the fish sinks to the seabed, settling on one side only, some of the fins changing position, and one eye literally moving round to join the other on the upper side. Other species of flatfish, such as the sole, halibut and turbot, can be identified by their varying sizes and patterns, but all possess similar habits.

Three-spined stickleback Eyes that lack lustre or expression, no matter to whom they belong, are often described as fishy eyes. In speaking this way we do less than justice to fishes in general. Take the three-spined stickleback, up to four ins long, of the freshwaters of Europe, North Africa and North America. The colour of the male becomes brighter as the breeding season approaches. His throat and underside become suffused with red, and the blue of his back and eyes becomes more intense.

The male stickleback builds a nest of plant strands stuck together with a secretion from his kidneys. When this is finished he watches for a female with a swollen belly, a sign she is ready to lay. He swims at her, showing his red breast, and leads her, by means of a stereotyped series of movements, to the nest where she lays her eggs. Several females lay in one nest and it is the male that tends the eggs and cares for the young.

Small diadematid sea urchin Sea urchins have few of the special sense organs so obvious in other higher invertebrates. Yet from their behaviour we can suspect they are sensitive to what is going on around them. Some sea urchins living in tropical seas have unusually long and slender spines. The animals themselves look like grotesque pincushions and anyone treading on them with bare feet can suffer painful wounds from their spines. Moreover, these sea urchins do more than rely on passive defence. When a larger animal that might eat them swims near, so that its shadow falls on them, they move their spines, thus directing them towards a possible enemy. The rows of blue eyes along the spines may have something to do with this: they are no more than sense-receptors capable of detecting shadows.

Peacock butterfly The peacock butterfly is one of the larger Eurasian butterflies, ranging from the British Isles to Japan. It is also one of the more showy when it settles on a flower and opens its wings, exposing their upper surfaces. The colours in the wings are mainly red, brown and yellow with some black. The conspicuous features are the large eye spots. The two on the hindwings look convincingly like true eyes.

It has long been believed that eye spots of this kind scare birds away. A few years ago this was put to the test. A box was placed in an aviary containing a resident jay. This had wooden sides and a glass top beneath which was an electric bulb that could be switched on and off. A mealworm was placed on the glass. The jay flew over, seized the mealworm in its bill and ate it. Then a photographic transparency of a peacock butterfly was placed between the light bulb and the glass. This time, when the jay flew over to take the mealworm, the light was switched on as it

lowered its beak. The effect was dramatic. The jay flew away, leaving the mealworm.

The experiment proved beyond doubt that when an insect suddenly exposes a pair of eye spots it frightens away an insect-eating bird. That it is the resemblance to a pair of eyes that matters can be proved by placing an actual butterfly's wing between plates of glass, after the blue scales of the 'eye' have been removed, leaving a blind eye. The birds are no longer scared by it.

Eyed hawk moth There can be few more remarkable ornamentations on the wings of insects than those known as eye spots. They look incredibly like real eyes, yet have nothing to do with vision. The eyed hawk moth of Europe has such a pair on its underwings. When the moth is at rest, the underwings are covered by the front pair of wings. When the moth is disturbed by an insect-eating bird it moves its front wings, exposing the underwings with their eye spots.

Experiments have shown that the sudden appearance of these 'eyes' startles a bird so that it flies away. The more the eye spots resemble true eyes the more the bird panicks.

Siamese cat The flowers known as forget-me-nots and the Siamese cat seen here among these flowers both owe their popularity, at least in part, to the colour blue. In the forget-me-not the blue is in the flower; in the Siamese cat it is in the eyes. Blue eyes are much admired in people, yet the reason why they are blue is that they lack the pigment in the iris which, in people with brown or other dark eyes, allows the green, yellow, orange and red rays to penetrate to the back of the iris. The blue is therefore reflected back.

Siamese cats are not fully albino (see Chapter Six) but have a temperature-sensitive albinism. Their ears and feet are cooler than the rest of the body, as is usual, and these are still pigmented. Nevertheless, the Siamese cat, like the white tiger and white rat, has an

abnormality of the optic nerves which impairs its vision. In the Siamese cat and white tiger this gives them cross eyes; and in both the eyes are blue.

Following page top:

Velella and sea snail Velella is a kind of blue jellyfish which is made up of a colony of polyps. One of the polyps is converted into a disc-shaped, gas-filled float with a triangular sail that sticks up out of the water. The wind catches the sail and drives Velella along. This has earned the jellyfish the common name of 'by-the-wind-sailor'. On the underside of the float are other polyps. Some form tentacles armed with stinging cells. Others catch small crustaceans and fish larvae. Velella's main enemy is the purple sea snail, which floats hanging to a raft of bubbles of its own making. In spite of Velella's stinging cells, the sea snail often eats all but the float and sail.

Velella and sea snail *see previous page.*

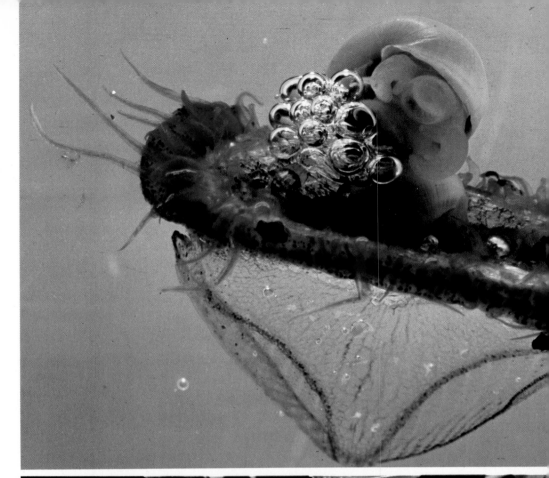

Blue gourami (Centre) The blue gourami, a freshwater fish of Southeast Asia, is often kept in aquaria. Another name for it is the three-spot gourami. In the wild it really does have three spots, one at the base of the tail, the other in the middle of each flank, the third on the gill cover. In the aquarium form the third spot tends to be lost. When a fish has a dark spot near the base of the tail, there is always the suspicion it uses it to confuse its enemy. Fishes sometimes swim backwards and then the spot near the tail may look like the eye.

Gouramis belong to a group known as labyrinth fishes. They have a special labyrinthine breathing apparatus in a cavity above each gill chamber. Each cavity contains a series of bony plates that support rosettes of membranes rich in blood vessels. These can take oxygen from the air. Labyrinth fishes can live in stagnant water by rising to the surface and gulping a bubble of air. They can also use their gills normally in water not deficient in oxygen.

Powder-blue surgeonfish (Below) Some coral-reef fishes are so curiously coloured that they look unreal, like painted fishes in a painted ocean. They look strange largely because we are so used to fishes having colours that make them look inconspicuous against their natural backgrounds. This powder-blue surgeonfish certainly shows a quite remarkable pattern of colours. Yet among the many coloured animals that make up a coral reef these patches of colour break up the fish's outline and the powder blue of the body takes on a neutral tint. Surgeonfishes are so named because they have a sharp spine, like the blade of a surgeon's lancet on either side at the base of the tail.

Blue-throated tree agama (Right) These lizards have short broad heads, and live in southern Asia, Africa and Australia. They are small to medium-sized reptiles, with powerful claws and a short, thick, slightly forked tongue. Agamas change colour according to the temperature of the surrounding air and their own emotional state. Some of them are especially noted for having bright colours around the head and neck which are used to frighten away rivals or enemies. The blue-throated tree agama of Africa is seen here in an aggressive display. It has seen a rival approaching its territory and is putting on what to another agama is a terrifying display, opening its mouth wide and showing the blue in its throat. If this intimidation does not have the desired effect, there will be a fierce battle between the rivals, each agama snapping at the other's throat until one of them admits defeat and slinks away.

Yellow long-nose butterflyfish The yellow long-nose butterflyfish is one of many fishes with gaily coloured bodies, flattened from side to side, that flutter around coral heads. They probe the crevices of the coral for small animals. At the first sign of danger a butterflyfish dives quickly into the spaces in the coral, out of sight. It seems therefore not to use colour as camouflage. It is, however, strongly territorial and seems to use its colours to warn off intruders. The one shown here raises its dorsal fin into a crest at the same time to increase the threat signal to a trespasser in its territory, like a dog raising its hackles.

We are accustomed to the idea, through the many studies on birds, of animals occupying a territory throughout the breeding season. Some butterflyfishes do not hold their territories all the time. In some places on a coral reef one species will be holding territories by day and at night go elsewhere to sleep. As they do so, another species, which has been resting by day, moves in to take over the territories for the night.

Yellow - life giver or death warning

The most universally familiar yellow object is the sun, visible in all parts of the world. The sun guarantees the continuation of life. Plants cannot live without it, and without plants there would be no animals. Even deep-sea animals living in eternal darkness depend on it utterly. Their food consists of the bodies of dead plants and animals sinking down from the ocean's surface bathed by the sun. It is not surprising there have always been sun worshippers. Indeed, the golden-yellow sun is symbolic of life, and yellow could be the life colour.

Among insects whose bodies are richly ornamented with yellow are wasps. They sting, they are a menace, and they are recognized as a danger by man and beast. In them, the yellow is not pure, but is combined with black. This combination of black and yellow is one of several that have earned the name of warning colours. They warn other animals that the insect, snake or whatever is wearing them carries a sting or contains a poison, or is unpalatable in other ways. Red and yellow both occur as warning colours, alone or in combination, often, even usually, associated with black (see also Chapter 8). But yellow is the most frequent colour of these danger-signalling combinations.

How important warning colours can be is shown by the number of animals, themselves harmless, that mimic the colours of harmful or unpleasant animals. An example is the wasp beetle that mimics the wasp.

Wholly yellow animals are not too common, indeed they are relatively few among the million or so known species in the animal kingdom. Prominent among them are two popular domestic pets – the canary and the goldfish. Perhaps men were influenced in the first place to use gold as a precious metal because it matched the colour of the sun. It may be that subconsciously we find yellow or golden pet animals attractive for the same reason.

There are plenty of other animals that include yellow, in spots or patches, as part of the general pattern of colour on their bodies. Without going into particular detail, it may be said that these spots and patches help as camouflage, aid recognition by animals of their own species, and are a spur to courtship and mating.

How then does an animal make the distinction between yellow, whether golden yellow or orange, that is not a warning of danger, and yellow that is a signal of danger? The answer seems fairly obvious. It is when the colours, yellow or red, are bold, well defined and well contrasted. That is where the use of black is so important, for it throws up the other colours and makes them catch the eye. When seen on its own, or not in association with black, yellow looks much softer.

Many flowers are yellow, or have yellow centres, and animals need some safeguards against confusing these with unpleasant or dangerous species. For that matter, the yellow tints of autumn in temperate latitudes could provide many problems.

Happily, 'All that glisters is not gold', and not all that is gold or yellow signals danger to the eyes of animals. This only occurs when the colour is emphasized by the presence of black. Thus are animals spared having to decide when yellow or any of its variations spells danger or is without menace.

Yellow is a pigmentary colour widespread in animals. It is responsible not only for the yellow colour itself but for much of the green, by combination with the structural colour, Tyndall blue.

53

Ladybird The ladybird is probably the best loved of all insects. The reason for its popularity may perhaps be its smooth rounded shape and bright colour, combined with its harmless habits; or it may solely be due to a human whim. Certainly the ladybird is useful to us since, both as a larva and an adult, it feeds on the damaging plant lice known as aphides. Most ladybirds are red with black spots. Some are black with red spots and others are yellow with black spots. All are warning colours that indicate to birds that these insects are unpalatable. This picture shows a seven-spot ladybird that has just emerged from the pupa. It is then yellow: the red-orange colour and black spots do not appear until later (see page 127).

The yellow pigment is called a lipochrome, which is a name loosely applied to several pigments responsible for yellow, orange and red. The darkening of the yellow to orange or red in the ladybird, and the appearance of the black spots, is due to the laying down in the cuticle of melanin. In some insects it is known that a diet deficiency can inhibit the formation of melanin. They are then abnormally coloured. Presumably, therefore, the newly emerged ladybird must feed to reach its normal adult colouring.

Anole lizard Most lizards are so coloured that they harmonize with their habitat and are not easily seen. Anole lizards of the Americas are no exception. They live in forests or in gardens among trees and shrubs and their green bodies get lost among the leaves. There are many races of anole lizard, all green and the males all have a differently coloured dewlap. This cannot be seen until one male trespasses on the territory of another. The owner of the territory thrusts out his dewlap, which may be yellow or orange, in an effort to intimidate his rival.

This male anole lizard of Barbados is displaying aggressively to its own reflection. Birds are well known for attacking their own image in the glass of a window, in the breeding season, reacting to it as they would to another male. Anole lizards are content to make threatening signs, such as extending to the full the coloured dewlap. The lizard in our picture was kept as a pet in the house. A mirror stood on the floor, leaning against a wall, and from time to time the anole would drop what it was doing to wander back to this particular room to remind his 'rival' he was still around.

Tree frog Frogs and toads were, so far as we know, the first animals to have a true voice. There are animals lower in the scale that can make sounds, such as cicadas and many fishes, but these have no vocal cords. Frogs and toads use their voices mainly during the breeding season and they do so by passing air backwards and forwards across the vocal cords, with the mouth shut. So they can still call when underwater. Sometimes a frog will scream when caught by a predator. That is done with the mouth open. The mating call is, in many species of frog and toad, made louder by the use of air sacs. These give resonance. There may be a pair of vocal sacs, one either side of the head, or one large one under the throat, as in the East African tree frog shown here. When the frog ceases to call, the inflated sac subsides.

Many frogs and toads merely give an unmusical croak but there are some that produce remarkably musical sounds. The green house frog of the Caribbean introduced into the United States whistles like a cheeping duckling, the Mexican burrowing toad makes a cry like that of a bird, the Cuban tree frog snores. Other frog calls can sound like the popping of corks, the crowing of a rooster, the bleating of sheep, the sound of distant cowbells or the tinkling of sleighbells.

Brimstone canary The canary, originally called canary bird, was first imported into Europe as a cage bird in the sixteenth century. It is a finch native to the Canary Islands, as well as Madeira and the Azores, islands in the eastern Atlantic. In the wild state it is greenish yellow with dark streaks and a greenish yellow breast and rump. For a long time this was the colour of the caged birds, which were prized for their song alone and fetched very high prices. Then, instead of importing the birds, people took to breeding them. The first result was an all-yellow canary, which gave its name to a colour, canary yellow. All that had happened was that a canary had been born that lacked the black pigment, melanin, so that the yellow was undiluted.

There are 31 close relatives of the canary bird in Europe and Africa. All are yellow and yellow-green in various proportions, according to the amount of melanin and its location in the plumage. Most are called serins, after the serin finch of central Europe, which is coloured like the wild canary bird. There is, however, a serin living in East Africa which has been given the name brimstone canary because of its sulphur-yellow front.

Flat periwinkle Periwinkles are sea snails. Like their relatives, the land snails, they feed on plant matter, in their case seaweeds. Most periwinkles have shells that are dull in colour so that they do not show up against a rocky surface when the tide goes out and leaves them high and dry, provided they remain still. The flat periwinkle of European coasts is an exception, usually being bright yellow or orange. There is, however, a wide variation in the colour of the shell. Although commonly plain yellow, it may be orange, brown, black, banded yellow and black, orange and black or brown and black. Sometimes all these colours and combinations will be found on a single patch of weed. It has been suggested that the shape of the flat periwinkle's shell gives protection because it looks like the bladders of the seaweed on which this species feeds. This seems unlikely since, in any case, the colours of the shell make the periwinkles conspicuous.

Yellow seahorse Sea horses live in warm shallow seas in many parts of the world. Looking like the knights in chess, they can cling to seaweeds with their prehensile tails. They can also climb among the seaweeds in which they live, using the chin. They have a tubular mouth which sucks in small animals from the plankton; and since the eyes can be swivelled independently of each other, they can keep all the keener watch for their prey. They are peculiar also in that the female lays her eggs in a pouch on the male's abdomen. There they are fertilized and develop, and later the young are born, the male ejecting them one at a time by muscular effort. Sea horses can also show remarkable although slow colour changes. Some species can range from white to yellow, brown or green, bright red, blue and deep purple. The group shown here are from the Indo-Pacific.

Not a great deal is known about these colour changes but if a sea horse is disturbed it loses its bright colour and goes brown. For example, aquarists find that if they transfer these fishes from one aquarium to another, they go brown and only slowly regain their bright colour, perhaps taking as much as three weeks to do so.

Platy Platies are small freshwater fishes that bear live young. They are found from British Honduras to Mexico. In the wild state this 2½-in fish is usually olive green with a pair of black spots near the tail. It is, however, variable in colour, from red to black. Platies are popular with aquarists and many more colour varieties have been bred in captivity, especially yellow, orange and red. These are known to aquarists as yellows. Others have much blue on their bodies which are yellow with a red tail, and these are called blues. As is common in the animal kingdom, it is the males that are dressed in bright colours. Female platies are, by comparison, dull.

Golden fishes A number of freshwater fishes show golden tints or have golden varieties. Several of them have been bred in captivity and selected to give a more intense golden colour. The goldfish of Asia has been selectively bred for centuries to this end. In the group shown below are also the golden tench, golden rudd, golden orfe and golden carp of Europe.

There is some variation in colour in all animal species, even in the wild. This is more marked in some species than others. The breeder takes advantage of this. When the Chinese first started keeping goldfish they doubtless found that every now and then a yellowish individual would be hatched. By putting these aside and breeding yellow with yellow the colour became intensified until finally there was produced the goldfish as we know it today. The same goes for the other golden fishes.

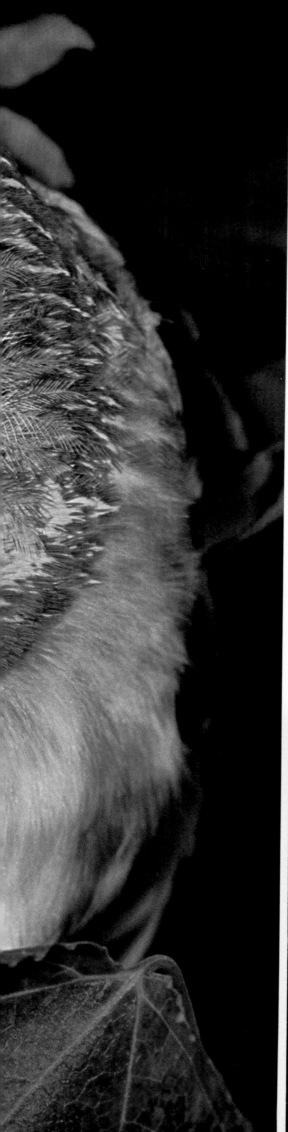

Saw-whet owl Many owls have golden-rimmed eyes like this saw-whet owl of North America. This is a common species from Alaska and California in the west across to eastern Canada and to the northeastern United States. It is a small owl, very vocal, with a voice resembling a grasshopper's song or a saw being sharpened. The saw-whet is seldom seen in full daylight but at twilight and during the night it hunts small mammals and birds up to its own size. It may also take bats and frogs.

The yellow iris of the saw-whet's eye, as in other owls, is most in evidence by day. At night the iris is drawn right back, exposing as much of the lens as possible. By day it closes to leave a very small aperture, in order to keep out light. The eyes of owls are 35–100 times more sensitive than ours; which is why owls can hunt by night or fly safely through a tangle of branches in darkness.

West African dwarf hedgehog It is tempting to compare the hedgehog, with its coat of spines, to the prickly plants of desert regions. Add to this its generally sandy colour, so typical of desert animals, and it seems reasonable to deduce that the dozen species of hedgehogs, living in various parts of Eurasia and Africa, are all descended from ancestors that were true desert animals. Perhaps the idea is not wholly fanciful, especially since half the present-day species live in deserts. In fact, colour varies from one species of hedgehog to another, from yellowish to brown, even to black. It is always sombre, as befits an animal that forages at night, mainly for insects, slugs and earthworms. By day, while it sleeps, its colour must, for the sake of security, match the ground or the dry grass and leaf litter. This is helped by the spines having light and dark bands, giving something near to a 'pepper-and-salt' effect. This, as we shall see in Chapter 8, is one of the best forms of camouflage.

Red-backed scrub robin Baby birds in the nest open their beaks wide when one of the parents comes in with food, even before their eyes are opened. They do this as an automatic response to the vibration of the nest as the parent lands on it or on the twig or branch supporting it. The nestlings' throats are always a vivid colour, usually yellow or red, and easy to see, so the parent need lose no time in looking for individual throats into which to push the food. These day-old nestlings of the red-backed scrub robin, of East Africa, are exceptions to the rule. They react by gaping in response to a sharp whistle.

There is another important effect of the vivid colour. In a nest, the hungriest nestling will gape widest and will thrust its head up the most in the direction of the parent's beak. So the parent does not need to remember which youngster it fed the last time. It simply pushes food into the most obvious and insistent throat.

Kongoni Hartebeests are unusual African antelopes, brown to yellow, up to 4 ft high at the shoulder, their backs sloping from shoulder to rump. They have long thin faces which look even longer because there is a hump on top of the head from which spring the bracket-shaped horns. There are several kinds. The one living in Kenya and Tanzania is called Coke s hartebeest; another name for it is kongoni. The young kongoni, shown here, is able to walk when about an hour old.

Kongonis inhabit the open grasslands or grasslands covered with sparse scrub. There the young kongoni needs its sharp eyes and keen hearing to give early warning of the approach of an enemy, of which it has many. Even so, its main protection lies in its tawny coat, and it is hard to see when crouched on the ground.

Lion The definition of some colours has changed through the centuries. Originally, tawny meant various shades of brown. Now it is used especially for brown tinged with yellow or orange. The one animal that acts almost as the yardstick for tawny is the lion. This superb beast formerly ranged across southern Asia and the whole of Africa. Now the realm of the King of Beasts is restricted mainly to East Africa, with a few hundred in a reserve in India.

Does a lion need camouflage? Normally it does not. It can take care of itself in a fight and its prey is more often than not warned of a lion's presence by catching its odour. Indeed, the prey can tell whether a lion is hungry or not by this means. A well fed lion can walk through a herd of zebra who will merely stop grazing to watch it pass. But lions often combine to spring an ambush, some of them driving the prey towards a few of their number lying hidden, masked by their tawny colour. That is when the tawny coat is most valuable.

Desert Rat or **Gerbils** These rats are found in desert regions of Asia Minor and northern Africa. They dig simple burrows, all with separate exits, in fairly soft sand. The burrows are usually situated in places partially sheltered by desert shrubs. The rodents forage for seeds, leaves, bulbs, stems and roots mainly at night, but sometimes they venture out during the day, when the countershading effect of the body might provide some protection against diurnal birds of prey. This is achieved by the whiteness of the underside of the body which lessens the effect of any shadow cast by its own body.

Hamsters In 1930 a nest of golden hamsters was dug out of the sand in Syria. It contained a mother and twelve babies. From these were bred others and in a very short time hamsters became popular pets, displacing the once favourite white mice. Today, there are millions of pet hamsters, all descended from this one family. The young are born blind, naked and helpless, but they have a voice and they use it, often. Even before their eyes are open they are active in the nest. At two weeks of age, when the eyes are open, they are weaned and are then almost wholly independent.

The golden hamster is only one of several related species in central and southeastern Europe. They are all much alike, except in size and colouring, and the exact relationships of one to another are not clear. One thing they have in common is that they feed on grain and other plants and have cheek pouches used for transporting food back to the nest. They put food into the mouth and work it into the cheek pouches. Back in the nest they push the food out, using their forepaws. One grey hamster, a European species, was seen to cram food into its pouches until its head and bulging cheeks made up a third of the animal's total volume.

Pale clouded yellow Until the present century, when studies of their migrations were made, little was known of the long distances travelled by some insects. The regular migrations of the monarch butterfly in North America were well-known and these compare with those made by birds from summer to winter quarters. Then it was found that certain butterflies in Europe and Africa made astonishing and puzzling migrations, puzzling because there seems no point in the journey. One of these is the pale clouded yellow butterfly, 2 ins across the spread wings. It breeds in North Africa and around the Mediterranean Sea and migrates each spring and summer, as far north as the British Isles. It used to be thought that these were journeys with no return, a sort of suicidal journey into the unknown. There is now reason to believe that some of these migrating butterflies do make the return journey, but it still leaves unsolved precisely *why* they make the journey.

Tiger swallowtail (Below left) This is one of the most familiar butterflies of North America, named for its yellow colour with dark stripes. It is found as far north as Hudson Bay, in Canada, and south as far as Florida and Texas. Its caterpillar is large, green and smooth with two large eye spots behind the head. It feeds on birch, poplar, ash and other trees such as cherry, plum and apple. In southern parts of its range as many as 50 per cent of the females may be brown.

It is believed that this dark southern form

is what is called a mimic of the pine-vine swallowtail. In the course of evolution it has come to resemble another species very closely. The idea is intriguing since the larva of the pine-vine swallowtail feeds on a poisonous creeper, *Aristolochia*, and retains the poison in its own body. This makes it distasteful to birds. The poison passes into the adult butterfly which is also distasteful and is avoided by birds. The mimic has no such poison but presumably is left alone by birds because it looks like the pine-vine butterfly.

Orange-tip butterfly The male of the orange-tip butterfly of Europe has white wings dappled with dark spots, especially on the undersides of the wings. He has, however, an orange patch at the tip of each forewing. The female is like him except that she lacks the orange tips. When she settles on a flower, especially on flowers of wild parsley, she seems to disappear, because of her dappling. So does the male, because he folds his wings in such a way that the dappled hindwings cover the orange patches. So he shows the orange only when flying, and then he flutters his wings so that the orange tips may more effectively catch the eyes of females.

Grass yellow butterfly (Below) These are small butterflies of tropical Africa, about an inch across the spread wings. They are found equally in rain forests and in dry scrub. Even in places where other butterflies are lacking the grass yellow is likely to be seen. Not

uncommonly they swarm on damp river edges, sipping the moisture. A similar and related butterfly, the least sulphur, is common over most of the eastern United States. Large swarms are often seen over the Atlantic and some have made the 600-mile crossing to Bermuda.

This incredible journey for so small an animal probably depends on favourable winds to assist it. One of the more recent discoveries is that butterflies and moths that take no food other than nectar must convert it into fat before it can be used as fuel for flying. Flapping flight consumes more energy than any other animal activity. Insects using fat, however, consume only about one per cent of their body weight in an hour as compared with flies, that use carbohydrate, which may consume as much as 35 per cent.

Purple heron Green seems the most effective colour for hiding among grass or other vegetation, but it is not the only one. Under certain circumstances of colour pattern and behaviour, an animal without trace of green can apparently disappear against a green background. A good example of this is seen in the purple heron of Europe and Africa. By virtue of its markings and the posture it adopts when alarmed it is very hard to detect. It is a sort of optical illusion.

The purple heron, pictured here in a papyrus swamp, illustrates the principle that the colour of an animal when seen at close quarters does not necessarily explain the camouflage effect produced by that colour. For example, the green woodpecker of Europe, which measures a foot long, is green with a yellow rump and a red crest, and stands out clearly on grass-covered ground at a distance of 30 yds, even though it is green against green. It is very obvious, also, when it flies up at one's approach and flashes its yellow rump. A green woodpecker launching itself from a tree, as well as during actual flight, is equally conspicuous. Yet when it lands on a tree it perches vertically on the trunk and seems to disappear although it is a green object on a brownish-black background. Even from as close as 10 yds it is practically invisible.

Green as grass and leaves

One great difference between a town and the countryside is that in the countryside there is a great deal of greenery. In the town green forms only a small part of the landscape. Since so much of the countryside, whether we are talking about cultivated pastures or tropical rain forests, is green, we should expect that, if camouflage is to be used, green should be a predominant colour in it.

In practice green is found almost entirely in land animals. It is confined mainly to small animals, especially those that do not fly. Green caterpillars are fairly common, whereas green moths and butterflies are rare. Also, green is not a common colour in birds, apart from some that live in dense forests.

Green is, however, a common colour among reptiles and amphibians. These are animals which, for the most part, remain still for long periods of time and, when they do move, cover the ground quickly by running, leaping or hopping. We can say of a frog, for example, that it only moves quickly in short bursts, as when flicking out its tongue to catch an insect, or when going from one place to another. Then it makes the journey in a short space of time by a series of jumps, each of which covers a long distance for a creature of its size.

More or less the same rule applies to insects. Green grasshoppers remain motionless, feeding or resting. To get from one place to another they take remarkable leaps for their size, and then stay still again. The great green grasshopper of Europe, is a good example. It is nearly 3 ins long. It perches in bushes, and it makes tremendous leaps. The difficulty is to find it, and if the grasshopper did not sing this would be even more difficult.

A uniform green colour over the whole body would often be a disadvantage to animals that cover distances at speed. It would be fine until the animal moved; but then the moving block of green would give the game away. This condition is most nearly achieved, among birds, in the green broadbill. Interestingly enough, this is yet another animal that stays still for long periods of time and then moves swiftly.

More commonly the green is broken up by stripes of paler or darker green, making the animal's body look more nearly like the vegetation among which it lives. Plants and the separate parts of plants are seldom wholly monochromatic, that is, of a uniform single colour. And this is the scheme we find, when we look more closely into it, in green animals. More commonly, however, green is most effective when used in spots and blotches.

The brilliant green so characteristic of lizards, snakes and frogs is due to three layers in the skin. There is an outer layer of yellow pigment beneath which is a Tyndall blue which consists, as we have seen in the introduction to Chapter 3, of a scattering layer against a background of melanin. Light passes through the yellow pigment to the Tyndall blue and is reflected back through the yellow.

The green in some birds, for example in many Australian parrots, has the same origin. Wild budgerigars are mainly green. Many colour varieties have been bred in captivity. The blue variety has lost its yellow pigment; the yellow variety lacks melanin; and the white budgerigar lacks both the yellow pigment and the Tyndall blue.

When specimens are preserved in alcohol in museum collections, the yellow pigment is dissolved and frogs, lizards and snakes that were green in life turn blue. In the green magpies of Southeast Asia the yellow pigment tends to fade in strong light. The birds turn blue if they leave the dense forest and stay for a while in the thinner forest margins, in the sunlight.

Emerald boa Boas are tropical South American snakes that kill by constriction. They throw their coils around prey and as it breathes out the coils tighten their grip until the prey animal cannot breathe. They do not crush their prey, as is often said. Related to pythons, they are believed to be the oldest of existing snakes. The emerald boa, or tree boa, up to 6 ft long, lives in trees, holding on to a branch with its prehensile tail, leaving the forward part of the body free to deal with passing prey, in this instance birds, squirrels and lizards. When resting it coils on top of a branch with the front part of the body inside the outer rings, looking like a bunch of green bananas. Its bright emerald green with creamy or white markings break up the outline so that the snake is almost impossible to detect as long as it remains still. Most other boas are predominantly yellow or brown, the colours being arranged in diamonds or irregular patches.

The young tree boas are very different from their parents. They are yellowish, sometimes pink, and their white markings are edged with dark purple or green.

Thick-legged flower beetle Although only a third of an inch long, the male thick-legged flower beetle catches the eye because it frequents flowers against which its metallic green is conspicuous. The reason that this European species is so named is that it has what can be described as balloon-like thighs on the last of the three pairs of legs. What purpose these serve is not known. They are not found in the female, which is smaller and more slenderly built than the male. The larva of this beetle lives in plant stems.

Soft-shelled turtle (Below) A young spiny soft-shelled turtle of North America, so called for the spine-like tubercles around the front edge of its shell, feeds on crayfish and water insects and is an example of change of colour with age. The adult is greyish olive with dark spots that are sometimes circles, or the shell may be blotched or mottled with light and dark patches. The young is more greenish and the circles are more clearly marked.

Turtles are reptiles with the body contained in a bony box, the outside of which is covered with a layer of horny plates. The upper part of the box is known as the

carapace, the underpart as the plastron. In most species both parts are rigid and strong. There are, however, soft-shelled turtles in North America, Africa and southern Asia. In these the bone in the shell is much reduced, there is no horny layer and the animals' armour is no more than a leathery skin. They seldom leave the water and to breathe they push their tubular snout through the surface to take in air. Moreover, the wall of the gullet is rich in blood-vessels so it acts as a gill, taking oxygen from the water, and turtles can remain at the bottom for hours.

Tuxalis grasshopper Green colouring is on its own almost sufficient camouflage for an insect living on vegetation with broad leaves. But additional protection is needed for a species that lives on grasses, and the tuxalis grasshopper of Africa has this in good measure. Its body is long and thin, like the leaves and stems it inhabits. The body is continued into a long, slender head and long antennae. The insect itself normally takes up a lengthwise position on the grass and the shades of colour on its body and legs match closely those found in grass foliage. *Cont.*

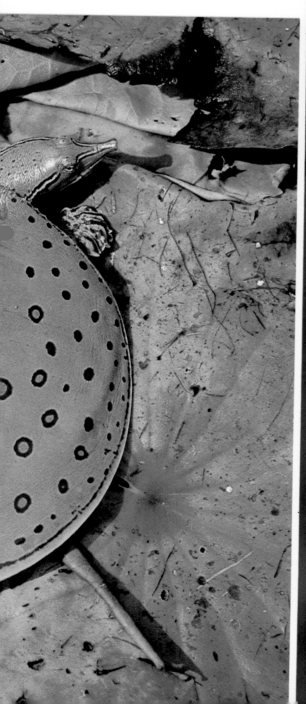

grasshoppers. The first were called short-horned grasshoppers, the second were long-horned grasshoppers, with short and long antennae respectively. Since short-horned grasshoppers are more closely related to crickets than to the long-horned type, their name was changed to bush cricket.

The chirping sound so characteristic of crickets is made by rubbing the rough surface of one forewing against the lower edge of the other forewing. Different sounds can be produced, some to attract a mate, some during courtship and others to repel rival males.

Green aphides (Left) One of the insect pests, particularly troublesome on cultivated plants, are the greenfly or aphides, also called plant lice. There are about 4,000 species of aphides, ranging in colour from white and grey through yellow, brown and red to black, but the ones that cling in colonies to the stems of our garden roses are green. They belong to the group of insects classified as bugs, which means they feed by sucking, using a beak or proboscis driven into the skin of plants. By thus sucking sap they weaken a plant, especially when present in large numbers. They also transmit viruses to the plants. Aphides are mostly wingless. The occasional winged individuals account for their spreading from plant to plant. Throughout spring and summer all aphides are females, giving birth by parthenogenesis, or 'virgin birth', to females until autumn when they lay eggs from which hatch males and females. These mate and the females lay eggs that hatch the following spring.

As we saw in the previous chapter, one of the reasons for the popularity of ladybirds is that they help to control the enormous populations of aphides. Chemical insecticides are also used for controlling them.

Costa Rican quetzal The male resplendent quetzal has been justifiably named as the most showy bird in the Americas and one of the most beautiful birds in the world. The female is slightly duller than the male but is still handsome. The plumage of the male is a bright metallic green, including the very long tail coverts. Such other colour as is present is hidden. It takes the form of crimson on the lower belly and under the tail. The resplendent quetzal lives in humid forests of Central America. It is the national bird of Guatemala and was revered by the Aztecs and Mayas as the God of the Air. Only the nobility were allowed to wear its feathers.

Quetzals live in humid forests, nesting in hollows, in trees, in an abandoned woodpecker nest or in one the quetzals have hollowed out themselves. The male shares the incubation and to do so must accommodate his long train that is about three times the length of the rest of him. He enters the cavity and then turns to face the entrance so that his train is doubled forward over his back and protrudes through the opening. Since quetzals have two broods a year, by the end of the breeding season his magnificent train tends to be in a sorry state.

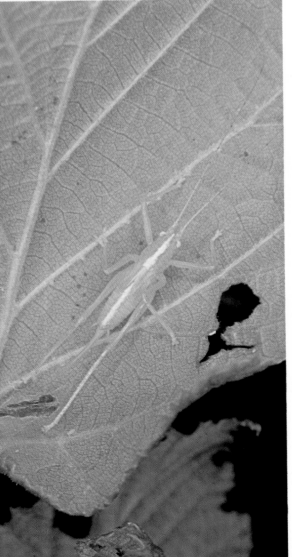

How is the green coloration of insects produced? Present in every living cell is an iron-producing pigment which plays an important part in cell respiration. It is called biliverdin because it is the green pigment in the bile of birds and mammals, where it is formed from the breakdown of hæmoglobin, the red pigment of blood. Biliverdin in its pure state is blue-green and in insects it is found either on its own or mixed with a yellow pigment. In grasshoppers, for example, biliverdin and a yellow pigment are laid down in the cells of the skin scattered in separate granules. They not only produce the colour green but the proportions in which the two pigments occur result in different shades of green, yellowish green or yellow, according to whether the biliverdin or the yellow is dominant.

Oak bush cricket (Left) The female oak bush cricket of Europe lays her eggs in crevices in bark or under lichens. She sometimes lays them in oak galls. It is not easy to see her doing this because this species is very strictly nocturnal. During the day the oak bush cricket stays on the underside of a leaf, beautifully camouflaged. Not only is the body a leaf-green colour but there is a pale green stripe along the back comparable with the pale green of a leaf vein. As if aware of the excellence of its camouflage, this bush cricket almost has to be forced to leave its hideout.

The name 'bush cricket' is quite new. Thirty years ago there were two kinds of

Malayan damselfly Damselflies are large insects with a wingspan of up to 4½ ins, related to dragonflies but more slender. The majority of them rest with their wings elegantly folded over their back. Many are brightly coloured in blue, red or metallic green; alternatively they may be colourless and transparent. The females usually lay their eggs in water where the larval stage is passed. This Malayan damselfly has transparent front wings and green hindwings. Like the male dragonfly, a male damselfly marks out a territory from which it drives out males of its own species. The Malayan damselfly advertises the boundaries of its territory by flashing its green wings when it lands.

Broad-banded rana Green is a common colour among frogs and toads. It is particularly common in tree frogs. Both these kinds of animal habitually remain still, and when they do move they move quickly. The broad-banded rana or Mascarene frog has green stripes and bands that break up the outline of the body producing an appearance of fresh grass blades against a background of old grass or earth. This is especially achieved by the green band along the middle of the back. In a photograph the frog is fairly obvious but in real life the frog shown here would readily escape notice until it moved.

The broad-banded rana lives in southeastern Africa near marshes and waterlily pans. The female is 1¾ ins long. The male is 1¼ ins long and makes a call like that of a Muscovy duck. Despite its excellent camouflage this frog takes no chances. When disturbed it leaps for the nearest water and either immediately leaps out again and disappears into the grass or goes to the bottom and stirs up the mud, effectively concealing itself as with a smokescreen.

Green mamba In Africa, the name 'mamba' is used for two snakes, the green and the black. Although related, they are very different in temperament and the bad reputation mambas have is due to the attacking habits and powerful venom of the ground-living black mamba. The green mamba, up to 7 ft long, is slender, and it lives almost entirely in trees in the forest or thick bush. Although it is venomous, with huge fangs in the front of the upper jaw, its aim is to avoid a fight and with humans it is shy and elusive and not particularly aggressive. It feeds on birds and their eggs and on tree-dwelling reptiles, such as chameleons and geckos.

There are two species, the green Congo mamba and the green Guinea mamba. The green mamba is not always green. It is bluish green at birth. Then, after the first sloughing (shedding of the skin) the head becomes dark green. At each sloughing this green extends farther back until, when the snake is about 2½ ft long, it is a uniform green throughout. Just before sloughing the skin turns a dark bluish-green or, in old individuals, almost steel blue. Surprisingly, the young of the black mamba are born green, for they too live in trees until they are large enough to survive on the ground.

Rajah Brooke's birdwing The Rajah Brooke's birdwing is the most striking and handsome of Malayan butterflies. It is perhaps the most beautiful of the several kinds of butterflies known as birdwings because of their size. They may have a wingspan of 10 ins or more. There are two races of Rajah Brooke's birdwing. The first is often seen in large numbers at seepages on river banks, drinking. Most of these are males, the females generally being hidden from view because of their habit of flying around the tops of tall trees. The other race is seen on flowers and there are as many females in evidence as males.

The males are slightly smaller than the females and are more beautifully coloured. The females are speckled with white. Nobody knows whether the black bands of the male

have any unusual meaning. They might be part of a warning coloration but so little is known about the habits of this and other birdwings that we can only guess about them.

Chameleon If all that is said or written about chameleons were true they would be the world's greatest colour-change artists. The usual statement is that they can change colour to match any surroundings – except a Scottish tartan. Chameleons active by day are usually green or bark-coloured. They go pale at night or if shaded from light by day. If a chameleon has been stationary for a while, with part of its body covered by a leaf, it will walk away with a pale yellow patch which is the exact shape of the leaf. A high

temperature makes a chameleon turn a lighter colour; low temperatures make it go darker. Their brighter colour changes are reserved for occasions when they have to impress their fellow chameleons, as when defending a territory. This applies also to the females of some species. The holding of territory is so important to a chameleon that it is impossible to keep a male pet chameleon within sight of another male of the same species. The colour change is therefore not a deliberate attempt at camouflage, although it does enable these reptiles to blend most effectively with the forest foliage.

Approximately half of the world's chameleons live on the island of Madagascar, the rest – apart from one Mediterranean species and three Asiatic species – being distributed throughout Africa south of the Sahara.

Green hairstreak (Below) The green hairstreak butterfly of Britain and continental Europe is less than an inch across. It is green on its undersurface, of both body and wings, and brown on the upper surface. When it takes off to fly it looks like a dead leaf in the wind. The resemblance is enhanced by the way it twists and turns in the air, sometimes looping the loop. When it settles and closes its wings, so that only the green undersurface is seen, it looks remarkably like a green leaf. It has the habit of returning time and again during the day to one particular shrub or bush and to one favourite spot on it.

Day gecko (Right) This brilliant gecko lives in the forests of eastern Madagascar, where it feeds on insects. The vivid colour is in keeping with its habits, for whereas other geckos are active at night this species, with rounded pupils adjusted to bright light, hunts during the daytime. It may also enter houses in search of its prey. Geckos are a kind of insect-eating lizard, usually brown or grey, living throughout the warmer latitudes of the world, even on oceanic islands. The 700 species are noted for their climbing abilities, loud voices and nocturnal habits. They have friction pads on the undersides of their spatulate toes, which enable them to run with speed up smooth vertical surfaces or scamper along a ceiling upside-down. The friction pads vary from species to species. In some geckos they are furnished with numerous tiny hook-like bristles that fasten to the minutest irregularities of a surface which may appear perfectly smooth to our eyes. In others the bristles end in minute suckers. Geckos have a peculiar wriggling gait. Each time they raise a foot they must curl each toe upwards, starting at the front of the toe, to disengage the bristles and allow the foot to be put forward.

Green broadbill In the tropical forests of Southeast Asia, from Burma to Borneo, lives the lesser green broadbill. It is a small bird coloured glossy green, except for a few black markings. Its head is large, its legs and tail short, and it frequents mainly evergreen forests. Believed to be a fruit-eater, it is remarkably sluggish for a bird, squatting close on its perch. When it does move, it hops swiftly along the branches or flies rapidly, usually for short distances only. Its leaf-green plumage harmonizes remarkably well with the dense foliage in which it lives, and its general reluctance to move conceals it all the more effectively.

Another remarkable feature is the lesser green broadbill's nest. Made of twigs, leaves and other vegetable matter, the nest is large and bag-shaped. It is suspended from a branch overhanging a stream by a long strand of fibres and it tapers below into an untidy tail. The result is that it looks less like a nest than a mass of flood debris that can be found hanging from the branches in such places.

Citrus swallowtail caterpillar When the caterpillar of the African citrus swallowtail butterfly is first hatched it looks like a small bird dropping. Because of this it is not molested when at its most vulnerable. When it is half-grown it takes on the appearance of a normal caterpillar. Then it is protected by two things. One is its green colour with black markings that break up the outline of the body, making it less easy to see than if it were a uniform green. The second, found in other swallowtail butterflies, is that it has an orange-coloured organ, forked at the tip, lying in its head. It pushes this out from time to time, to give off an unpleasant smell. This caterpillar feeds, in its native haunts, on the leaves of the orange. Here it is pictured feeding on a bush of mock orange in a greenhouse.

Although swallowtails are present all over the world, the majority of species is found in the tropics. Most, but not all, are distinguished by the long tails on the hind wings.

Giant clam Giant clams of the Indian and south-west Pacific oceans may be 3 ft across and weigh 300 lb. Their fluted shell, wedged in a crevice in a coral reef, opens to reveal a blue-green flesh. The giant feeds like other clams, by filtering small organic particles from water drawn in through its siphon. It also has an auxiliary source of food. Living in its tissues are single-celled green algae. As they die they break up and are digested by special cells in the clam's body.

Plants, whether they are microscopic algae or giant trees, need sunlight to manufacture food. They grow better when aided by artificial means, as when grown in a greenhouse or under cloches. The giant clam is an animal which works on a similar principle and has been described as the first cloche gardener. In its skin are small transparent areas that allow the ultraviolet rays of the sun to pass unimpeded to the algae growing in the tissues beneath. This increases the rate of growth of the algae, ensuring an adequate food supply to the clam.

Cuckoo wasp Cuckoos are not the only birds to lay their eggs in another's nest. There are others, such as the cowbirds of America or the honeyguides of Africa and Asia. There are insects that play this trick and one kind is the cuckoo wasp of East Africa. Only a third of an inch long, it looks like a brilliant metallic-green bee. The female lays her eggs in the nests of other wasps and leaves the hosts to rear the grubs that hatch from them.

There are some 1,000 species of cuckoo wasp in various parts of the world, all with bright metallic colours, usually green, sometimes green and ruby or blue. One peculiarity of their shape is that the abdomen is no longer than the head and thorax combined. This contrasts with the usual wasp shape, in which the abdomen is much the longer. The undersurface of the abdomen is flat or concave, and if disturbed the cuckoo wasp turns its abdomen under the thorax and rolls itself into a ball, leaving only the wings exposed.

Varying hare In winter the varying hare of North America has hairy soles that keep its feet from sinking too deeply into the snow. So another name for it is snowshoe rabbit. It is, however, a true hare, called varying because it has a brown coat in summer, which changes to white for the winter. In between it shows patches of brown and white or has a blue-grey tinge. There are, however, several varieties. The varying hare that lives in Arctic regions retains its white fur all through the year, only the eartips being black.

In Europe lives a near relative of the American varying hare, called the alpine hare. There are races of this species in the British Isles. One, in Scotland, is known as the varying or Scottish or blue hare. The one in Ireland, the Irish hare, does not turn white in winter with the same regularity as the others, the winter coat often being white with russet patches.

White and black and pied

White is not a colour but an absence of colour, a sort of non-colour. Yet it would be wrong to leave it out when considering colour. Although pure white animals are rare in the wild, they occur commonly in domesticated animals and among semi-domesticated animals kept as pets. White patches, spots and stripes are, by contrast, common and where these occur they mean much to the animals possessing them.

First, so far as wild animals are concerned, there are the species that are wholly white. They are mainly birds, such as most swans, herons and pelicans. There is also the white whale or beluga. Then there are those mainly white with some grey, as in some of the gulls.

Secondly, there are those abnormal individuals within a species that are wholly white, usually with pink eyes, known as albinos. Why one starling in a flock should be pure white is not fully understood. Nor do we know what the effect on the individual is, because albinos in the wild state are so rare that it is not easy to study the effect on them of their whiteness.

The complete absence of pigment in the skin, fur or feathers, that causes albinism is generally associated with some physical weakness. A pure white domestic cat is usually deaf. It may not be totally deaf when young but its hearing then goes as it grows up. Albinos are more common in some species than others, but in all they are rare, probably much less than one per cent.

White is most common in land animals living in the Arctic. A very few, the polar bear being one, are white the year round. Mostly there is a change from a coloured summer coat to a white winter coat, as in the Arctic fox and the ermine. There is a variety of the Arctic fox, known as the blue fox, which is smoky grey all its life.

It is usually said that a polar bear or Arctic fox, being white, can creep up on their prey unseen against the snow. It has also been said that a white coat retains the heat of the body better than a coloured, and especially a dark, coat. So a white winter coat is mainly for keeping warm.

Many animals are lighter on the underside than on the back or upperparts. This is especially true of fishes and birds. Many years ago it was suggested that this had the effect of making the body less obvious and of making it look less solid. This is known as the theory of countershading. An animal may be camouflaged in its natural surroundings but if you turn it over its white underparts stand out starkly. The theory seems reasonable and yet, in recent years, scientists have cast doubt on it.

Go to a large natural history museum, with hundreds of different animals on display. Take note of how many have a white underside. You will be surprised to find that a relatively small proportion are white or paler underneath. Those that have paler underparts are mainly fishes and water-birds.

It is impossible to go more fully here into the theory of countershading except to say that its weakness lies in the many exceptions. The whitish front of the European mistle thrush, for example, often acts like a mirror when the bird is

Polar bear It is often said that the polar bear's white coat enables it to creep up on its prey undetected against a background of snow. But there are flaws in the argument. For one thing, the animals hunted by the bear often have white coats themselves, and this should even things out. Secondly, a polar bear's coat is usually cream-coloured rather than pure white and thus tends to show up prominently against the snow. Furthermore, concealment is not possible during the summer, even in Arctic regions. The truth, as we now know, is that a white coat keeps in the heat, so preventing Arctic animals from freezing.

The polar bear is the most carnivorous and one of the largest of the bears. A well-grown male may be 8 ft long from its black neb (nose-tip) to its stump of a tail and weigh 1,600 lb. The female is only half this weight. The neck is long compared with that of other bears, the limbs powerful and the feet broad with hairy soles. The polar bear's main prey is seals which it stalks by taking advantage of the cover afforded by hummocks of snow.

Grey seal pup Young grey seals are born with a white coat, which is later replaced by fur coloured more like that of the adult. Around the coasts of the British Isles this white coat makes the seal pup conspicuous as it lies on the beach or on land near the shore. Elsewhere, as in the Baltic, where grey seals are born on ice, it may help to camouflage them. Yet the only enemy of the grey seal is man, so camouflage is probably not the answer. Perhaps it helps keep the seal pup warm. We can only guess about this because in the Antarctic seal pups are born on snow with dark coats. Should there be a few days sunshine the snow melts beneath the pups, as it does under any dark object placed on it, because a dark object readily takes up heat. The pups are unable to climb out of the pit so formed in the deep snow and starve to death. But why a grey seal pup born into less cold conditions should be more in need of a white coat is hard to say.

on the ground and gives immediate identification of the species. Another animal with good countershading is the grey squirrel.

White patches are most useful as signals. An animal may use them to advertise that it has taken over a territory or to indicate danger. Nocturnal animals can recognize members of their own species in the dark by the pattern of white patches. The white stands out when all other colours are lost in the gloom. Land animals that depend on speed for safety, like deer and antelopes, flash white patches on their rear parts as they run away. On the pronghorn of North America the patches reflect the light like a mirror and other members of the herd can see them over a long distance. They therefore know there is danger about and so are alerted.

To go to the opposite extreme, completely black animals are rare. Black, like white, is a non-colour. Whereas white is due to an absence of pigment, black is the result of one particular pigment, melanin. This is the pigment that governs hair colour and the colour in the cuticle of insects. Where black animals do occur their total blackness may be a family trait, as in the crow family among birds; or it may be that only some individuals within a species are affected, as in the case of the black panther which is a black form of the leopard. In the second instance these individuals are referred to as melanistic forms; that is, they are individuals of a species which possess more than the usual amount of the pigment melanin.

Black and white therefore fall naturally together, and if completely black and completely white animals are comparatively uncommon, there are quite a number which are a mixture of both black and white, a pattern referred to as pied or piebald.

Bottlenosed dolphin Few people had heard of the bottlenosed dolphin until the late 1930s when the species was first put on show in the so-called Marine Studios in Florida. This was the world's first dolphinarium, consisting of huge concrete tanks filled with sea water. Since then the idea has spread and dolphinaria in various countries attract millions of visitors. The bottlenosed dolphin, intelligent, friendly and easily trained, is usually the star of the show, performing the most remarkable kinds of acrobatic feats. But in addition to providing entertainment these dolphins have revealed to scientists many secrets, not only of their own life habits, but also of other dolphins and related species such as porpoises and whales.

Dolphins, porpoises and whales are placed in an order on their own, known as the Cetacea. The one thing we can say with certainty of this order is that there is a minimum of colour among its members. Usually they are dark on the back and flanks and white on the belly. The dark of the back

may be slate blue, grey or black. Some cetaceans show patches of yellow, some are all-white, some all-black.

Cetacea are undoubtedly descended from ancestors that lived on land. What these creatures were like nobody knows. In taking wholly to water they found themselves free of enemies. So there was no further need for camouflage. Their eyesight became unimportant, so the need for colour in courtship and species recognition became far less. As a result, any colour their ancestors may have had just disappeared.

Willow ptarmigan This is the willow ptarmigan in the snow of Alaska. It is found throughout the Arctic and Subarctic of both North America and Eurasia. In summer it has a brownish plumage that blends with the earth and the vegetation, giving the bird some concealment. This changes to white for the winter, and the white plumage helps to hide the bird against a background of similar colour. The white feathers also help to prevent the escape of heat, so keeping the body warm. The changes in colour of the plumage are, however, largely a matter of camouflage. The finely barred and patterned brown plumage of summer blends with the heather-covered habitat. The white of the winter plumage matches the snow. During March to April and again in October to December, when the plumage is changing and the birds are partly brown, partly white, they are living in a landscape that is half rock and heather and half snow.

Related to the willow ptarmigan are the white-tailed ptarmigan, found in the mountains of North America, the red grouse of Ireland and Scotland, and the rock ptarmigan. The last of these, known as the rock ptarmigan in North America and simply as the ptarmigan in Britain, also changes plumage according to the season, turning white in winter.

Llama The llama of the mountains of South America is a member of the camel family. Baby camels are much lighter in colour than their parents. So are baby llamas, and often they have pure white coats. If they were still living wild, instead of being domesticated, the babies would probably be darker in colour, instead of having white coats that might draw the attention of birds and beasts of prey. All the same, young llamas can look after themselves well. They can get on their feet and prance about almost as soon as they are born, but are not weaned until six to twelve months old.

As a result of selective breeding, llamas with completely white coats have been produced; and most adults have large patches of white, mingled either with brown or black.

The original Indians of South America and later the Incas, tamed llamas, using them as beasts of burden and as providers of meat, hides and wool. When the Spaniards came they employed llamas as pack animals for carrying gold and silver over the Andes. Nowadays comparatively few are left in the wild but domesticated llamas are still used in the highlands for transport purposes. In proportion to their body weight they have as much strength and stamina as camels.

Tawny owl The tawny owl, or brown owl, is one of the commonest owls in Europe and Asia. It is also found in northwest Africa. Its name describes the colour of the adult, which is reddish brown or brown, somewhat mottled and with small white patches in the wings. It seldom comes out by day and a tawny owl resting, perched on a branch close to where it joins the tree trunk, is very hard to see. The nest is usually in a hollow tree and the white plumage of the nestlings must be an advantage for the parents seeking to feed their offspring in the dark recesses of the nest cavity, because white shows up in the dark. Tawny owls seem to be able to locate their own kind and recognize them, at night, despite the camouflage afforded by their plumage.

The greatest danger to a young tawny owl when it leaves the shelter of its parents comes from its slowness in learning how to feed itself. To this must be added the difficulty it may have in finding a territory to occupy. Tawny owls are evenly spread out and usually the young owl has to find a territory vacant because of the death of the owner. In the course of its quest it may have to pass through several occupied territories, the owners of which harass the young one, giving it little time to hunt.

Greater sulphur-crested cockatoo The parrot
family consists of a large number of highly
coloured birds found especially in South
America and Australia. The colours of some
of them are quite outstanding, yet one of the
foremost favourites is the greater sulphur-
crested cockatoo of Australia, white except
for its sulphur-coloured crest and lead-
coloured beak.

Like the majority of cockatoos, this species
assembles in enormous flocks. In the wild the
sight of thousands of these birds flying out at
daybreak for food and drink, then returning
at dusk to roost, is truly remarkable.

Cockatoos form a small subfamily found
only in Australia, Papua and adjacent islands.
They are large parrots with a long erectile
crest, mainly white except for a few yellow or
red markings, such as on the crest, and a
powerful bill capable of crushing hard nuts.
All have this white plumage except for the
black cockatoo and the dark-slate, almost
black, raven cockatoo. The whiteness of
white cockatoos is due to lack of pigment in
the plumage. The black cockatoos have an
excess of melanin.

Albino swordtail (Right) This freshwater
fish readily produces albinos under
aquarium conditions and these are
popular with aquarists. This species is one
that throws a high number of albinos. Only
the male has a 'sword', which is the elongated
lower lobe of the tail-fin. The females give
birth to live young. Sometimes a female,
having had up to a hundred babies will
change sex. The lower part of her tail-fin
grows long and once this, with other changes
within the body, is completed, this new male
can father up to a hundred more offspring.
The fish top left in the picture is a young male
growing a sword.

This change of sex is possible, although
unusual, in all vertebrates because early in
life they posses both male and female
reproductive organs. What normally happens
is that one set of organs develops at the
expense of the other and the individual
becomes either male or female. Whichever
way it goes the vestiges of the other organs
remain and are capable of becoming active
again to outstrip the first set of organs,
converting male to female or, more usually,
female to male.

White pelican The great lakes of the Rift
Valley of tropical Africa are a home to
multitudes of birds, including white pelicans
and lesser flamingos. Both are unusual birds,
in build and in feeding habits. The flamingo is
unusual because it lowers its head upside-
down in the water to sieve small food. The
pelican, a fish eater, is noteworthy for its
enormous bill with the huge sac of skin
beneath it, used as a sort of landing net for
scooping up fish.

The half-dozen species of pelicans are
spread around the world in warmer latitudes.
When adult, they are white with black or
dark feathers in the wings, and they show
yellow, orange and red colours on the bill,

pouch or bare skin of the face. In some
species the white plumage is suffused with
pink in the breeding season. The young
pelicans are brown and there is one unusual
species which is brown throughout life. This
is the brown pelican of America. It is
exceptional too in being a marine species,
catching fish by plunging from a great height
into the sea.

Mute swan The mute swan of northern
Europe is a most spectacular bird, so much so
that in Britain it has for centuries been
designated a Royal bird. At one time
all mute swans were the property of
the monarch or were owned by nobles who
had special permission from the monarch to
do so. This swan so seldom makes a sound
that legend had it that it never used its voice
until it was dying. Hence we speak of
someone, such as a famous actor, who appears
for the last time before retiring as making his
swan-song.

There are eight species of swan in the
world – the black swan of Australia, the
black-necked swan of South America, the
coscoroba swan of South America and five
white swans in the northern hemisphere.
There is no obvious reason why all but two
should be wholly white, apart from coloured
bills and legs. Nor can one provide a
satisfactory explanation as to the advantages
of being all-white. It may be that whiteness
creates an optical illusion. If you see a group
of horses in a field and one is white, that one
seems much larger than the black or chestnut
horses around it. Yet measuring them will
show that all are about the same size. Swans
are large birds and their whiteness makes
them look even bigger. This alone helps to
deter potential attackers.

gills. Under certain conditions the retarded larva may change into an adult, an ordinary salamander, greenish black in colour. The name 'axolotl' is Mexican for 'water sport', a suitable name for an extraordinary animal.

When an animal remains in the larval state all its life it is said to be neotenic. Neoteny is most common in newts and salamanders. Some species, such as the Texas blind salamander, the mudpuppies of North America and the olm of Europe are all neotenous, live permanently underwater and have feathery external gills throughout life.

Fiddler crab (Below) When the tide goes out on certain parts of tropical coasts the sandy or mud flats are alive with small crabs, known as fiddler crabs. They are so named because each male has one claw much larger than the other. It is in fact about as large as the body, as in this male photographed on the shore of Malaya. Each male continually jerks his large white claw which, catching the sun, stands out in contrast with the drab background. It is as if he were beckoning, but he is, in fact, signalling that he has occupied a territory and at the same time attracting the attention of a female.

One thing that has been noted is that in many, if not all species of fiddler crab, the signalling changes when a female approaches a male. He may wave his claw more rapidly or the claw may be opened and shut in time with the waving. In some species the large

claw and the first pair of legs are vibrated rapidly, or special dancing steps may be made. Whatever is done, and whether for territorial reasons or directed at the females, the fact of the large claw being white increases its effectiveness as a means of making signals.

Albino rabbit For a reason difficult to discover, white animals seem to be more favoured as pets than those that have normal colouring. White mice, which are albinos of the house mouse, are especially popular. So are white rabbits, albinos of the European wild rabbit.

In all species of animals there is always the chance of a white individual, or albino, turning up. They are, however, rare. It has been estimated that in the wild albinos occur at the low rate of one in 15,000 to one in 25,000. In some species it may be as low as one in a million. Even these usually have a lower survival rate than normal individuals if only because albinism is so often linked with another physical defect, usually with defective hearing or sight.

The term 'white elephant' is commonly used to describe something that is useless yet valued by the owner. This has its origin in the fact that in Thailand the albino or white elephant – a great rarity – is sacred and not allowed to be put out to work. Its skin is, in fact, light grey rather than pure white and the eyes, as in all true albinos, are pinkish.

Axolotl In certain lakes around Mexico City lives a salamander that often does not grow up. What happens is that the larva frequently remains undeveloped, yet is able to breed without turning into an adult. This larva has a long body and a long tail. A fin runs the length of body and tail, and there are feathery gills on each side of the head. The colour of the larva may be black or dark brown, but commonly the larva is albino, white with pink

Stoat Stoat or ermine? It depends where you live. In Britain it is called a stoat while it is alive. When its white winter coat with its black-tipped tail is turned into fur for wearing on royal or noble ceremonial occasions it is called ermine. In North America it is always called the ermine. But it is the same animal on both sides of the North Atlantic. Only in cold regions does the brown fur of the summer coat turn to the white of winter. In some places, where winters are not severe, the coat turns only partially white.

The changes from the summer to the winter coat takes at most three days. This is not sufficient time for an entirely new coat to grow and what has happened is that the white coat has been growing unseen beneath the brown coat. Then, when there is a sharp drop in temperature of the surrounding air, the hairs of the summer coat are quickly shed, revealing the white coat. The change back from the winter to the summer coat, however, does not take place so quickly.

Grey squirrel The grey squirrel looks grey all over except for its underparts, which are white. Some people do not take kindly to this squirrel because it steals their soft fruits, or because it kills nestling birds or eats the eggs. Others find its antics most engaging. There is no doubt that the squirrel looks particularly charming when it squats on its haunches facing us, so presenting its white front to our view.

In fact, the grey squirrel is not grey at all. Its silver-grey appearance is due to the hairs being banded along their lengths in white, black and brown. They are black at the base, then yellowish brown, then black again, ending in white tips. Its summer coat is browner than the winter coat. Pure black, that is melanistic, grey squirrels are not uncommon, nor are gradations between black and the normal grey, due to interbreeding between the black and the grey. Nevertheless, there is reason to believe that melanism in this species of squirrel and in others is more common in areas where the climate is moist. In melanistic grey squirrels the underparts are orange-brown instead of white.

Gerenuk (Right) The gerenuk is a long-necked antelope of East Africa that browses the leaves of shrubs and trees. It is also called Waller's gazelle and is famous for its habit of standing erect on its hindlegs to reach even higher foliage. Like so many other antelopes and deer, the gerenuk has a white patch around its tail. The underside of its tail is also white. While browsing contentedly it is constantly flicking the tail, showing the white. If it senses danger it dashes for cover and the last thing to be seen is the white patch disappearing. To other gerenuks this conveys a message. The white patch and the flicking tail are automatic signals one to the other, indication either that all is well or that there is danger in the offing.

There is little information to hand about the gerenuk's enemies. Presumably, any large carnivores would be liable to take the young, or even the adults. The tail signals might

catch the eye of such a predator, but it is just as likely that the carnivore will be confused when the white patch suddenly vanishes as its owner dives into cover.

Kori bustard The kori bustard is a large bird of the plains of East and South Africa, especially common in Kenya. The picture shows a cock kori bustard in northern Kenya with the mountains of the Matthews Range beyond. The male is mainly black and greyish buff, but as the breeding season approaches he puffs out clumps of white feathers. This display is intended to let other males know he has established a territory. It is also used in courtship. The two purposes are largely combined as the male kori bustard tours his territory in a stately walk, his tail held up to display the conspicuous white of fluffed under-tail coverts.

The male great bustard of Europe goes further. He seems almost to turn himself inside-out, fluffing out large areas of white feathers that at other times are hidden under his brown plumage. Since this could not only attract the attention of females but also of enemies, the bustard uses a compromise. Although he looks like a ball of white feathers when displaying, he holds this posture for a few seconds only, at most for two or three minutes.

Baby monkey During the months before it is born, the baby monkey, like most baby mammals, has been sheltered within the mother's body. The exceptions are in the platypus and spiny anteaters of Australasia, which lay eggs, and the pouched mammals, or marsupials. In the marsupials the baby is born at an early stage but it climbs into the mother's pouch as soon as it is born, and it shelters there for several months.

In many of the other mammals the babies leave the warmth of the mother's body with hair on their bodies and must be sheltered in a warm nest. Kittens and puppies, for example, are born in a nest where they keep warm by huddling close to their mother and to one another.

Monkeys have one baby at birth, rarely twins, so there are no brothers and sisters to provide warmth. A baby monkey, although it has a short coat of hair, must therefore cling closely to its mother. She also wraps her arms around it. The baby has a black coat which it exchanges for a reddish-yellow coat as it grows. All the males and females in the troop take an interest in the baby but as the colour of its coat changes their interest wanes. So the youngster gains independence unimpeded by too much attention.

Siamang Gibbons are monkey-like apes living in Southeast Asia, from Assam and Burma as far east as Borneo. There are seven species which range in colour from brown and buff or almost cream to black. The largest of all, which is black all over, is known as the siamang. A young siamang is shown in this picture.

The siamang is about 4½ ft high when standing erect and its arms are so long it holds them over its head when on the ground. Like all other gibbons it is more at home in the trees, swinging from branch to branch on its long arms. Its hands are long but the thumb is short, so the hands are used as hooks instead of grasping the branches.

Food consists of leaves and fruits but the siamang probably eats birds' eggs and small animals it finds in the trees. Carrying food is no problem. Although the hands are needed for swinging through the trees, food is firmly grasped with the toes.

Raven Two elephants and a fan-tailed raven near a waterhole in an African setting provide a study in black. As we have already seen (p. 36), elephants often take on the colour of the mud in which they wallow or the dust with which they bathe themselves.

The fan-tailed raven is smaller than the original raven that is so widely distributed throughout the northern hemisphere, in North America, Europe, most of Asia apart from India and Southeast Asia and northern Africa. The northern raven is 25 ins long with a 4-ft wingspan compared to the fan-tailed raven which is 18 ins long. Also, whereas the northern raven has a hoarse croak, the fan-tailed raven's call is a shrill falsetto call. Both species favour regions where there are rocky cliffs and gorges.

Ravens are symbolic of blackness. A very old belief about the raven in Europe *cont.*

was that the bird was apprehensive when its naked chicks first hatched in case they turned out not to have black feathers. So until their feathers grew the parents refused to feed them. That this legend goes back a long way is suggested by the verse in the Psalms of David: 'which giveth fodder unto the cattle and feedeth the young ravens that call on him'.

Coot Coots are black birds that live mainly on water, rarely taking to the air, although once airborne they fly strongly. They swim strongly, dive easily to search for food, and have few enemies, largely because they know how to defend themselves. One of their more spectacular tactics is to use a water barrage. When a hawk or a crow flies in to attack, several coots will sit on the surface of the water and, using their strong legs with their long, lobed toes, splash water into the air.

The chicks are black with red around the head. As they grow they lose this red, although the eye remains red. Instead, the adult coot has a white bill and this continues onto the forehead as a white frontal shield, which contrasts sharply with the black of the rest of the body. One effect of this is to make the bill look larger, always an advantage when facing an adversary.

As so often happens with black-plumaged

birds, in strong sunlight a blue-green or a purple sheen may appear on the feathers.

Mole Moles live underground in the northern hemisphere, in Eurasia and North America. They are beautifully adapted for this. The body is cylindrical, so that it passes easily through a narrow tunnel, like a cork through a bottle's neck. There are no earflaps to impede progress. The eyes are small, which is no drawback since the moles seldom see daylight. Moles can run forwards as easily as backwards. Their legs are short and the hands are broad, forming powerful shovels for scraping away earth as they tunnel.

Their fur is usually black or dark brown but it is likely that colour is of little importance. Since they live underground, camouflage is not needed. A few inches below the surface the soil remains at an even temperature no matter what is happening at the surface, and moles do not live in Arctic regions where the ground is permanently cold. So it seems the dark colour of their coat could have little effect in taking in heat.

How little colour matters to a mole is shown by the European mole in the picture. It has been found that in certain localities most of the moles, instead of having black coats, may be white, cinnamon, orange or cream.

Black molly Black mollies are freshwater fishes popular with aquarists. The species was given the scientific name *Mollienesia*, and aquarists have a habit of shortening such names. Mollies are a kind of toothcarp that bear live young. The wild forms live in the northwestern parts of South America. They are olive in colour but selective breeding in aquaria has produced a number of colour varieties including a melanistic or black variety, the black molly. Breeders used mollies which had large areas of black on the body and it was only after many generations that the first all-black individuals appeared. Such mollies give birth to completely black young.

The black molly is one of the short-finned varieties. Even more decorative is the sailfin molly, so named because of its large dorsal fin. Selective breeding has also produced another spectacular black variety, the midnight molly, a cross between the sailfin and the black molly.

A most amazing discovery was made by the Americans Professor Carl Hubbs and his wife while studying the Amazon molly. They found the females could produce young with no males near provided there were males of other species near. It seems that the development of the molly's eggs is triggered without the sperms of these stranger males contributing anything to the inheritance of the offspring. In other words, it is a virgin birth induced by an alien father with the offspring breeding true.

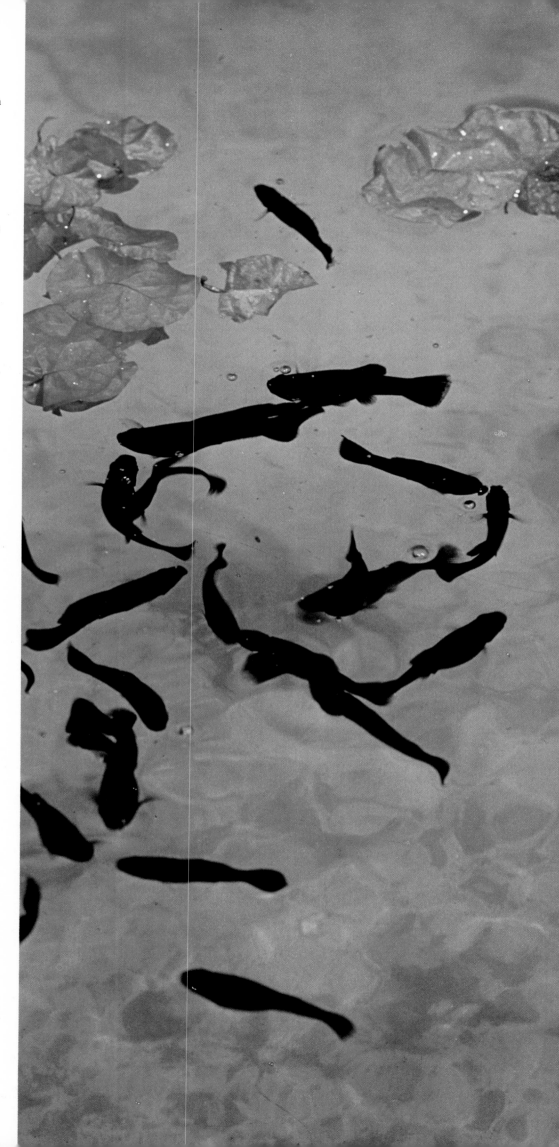

Giant panda The most celebrated of all pied animals is the giant panda, which is white except for black forequarters, hind legs, ears and circles round the eyes. It lives in the bamboo forests on inaccessible mountainsides in southeastern China and was unknown to Western scientists until 1869. Since then relatively few giant pandas have been seen outside China.

Its late discovery and apparent rarity would be sufficient to excite interest among scientists. Its popularity as a zoo animal is less easy to explain. In part it may be due to its human-style habits of sitting upright and holding food in its hands; and there are those endearing eyes, actually quite small but made to look large by virtue of the black patches surrounding them. All the same, it is extremely unlikely that these characteristics would have made it such a universal favourite if it were something other than pied.

What use is the black-and-white pattern to the panda itself? We can only guess because nobody has studied it in the wild. Probably this pattern makes it hard to detect in the dense bamboo thickets. This may be why it has so often been called a rare animal, although recent researches suggest it is not.

Penguin There are 17 species of penguins and all live in the southern hemisphere. A few live on the continent of Antarctica, others on subantarctic islands and on the coasts of South America, South Africa, Australia and New Zealand. The farthest north is the Galapagos penguin, on the equator. Right is a family of gentoo penguins; the adult shows the characteristic white eye patch.

With one exception – the little blue penguin of Australia and New Zealand – all have a white front and a black back. Penguins of each species probably recognize each other, just as we tell one species of penguin from another, mainly by the patterns on their heads, as in these chinstrap penguins. These are often black and white only but may include yellow and red tufts of feathers or patches on the face. The commonest of these two colours is some shade of red, from pink to orange or deep red, on the bill, on the feet, as a red rim round the eye or red eyes. Since black and white are non-colours we could say the dominant colour in penguins is red.

The emperor and king penguins make no nests. The female lays a single egg which she holds on her feet, protecting it with a fold of the skin of her abdomen. Others make a nest in a burrow or collect pebbles to form a crude nest as do the chinstrap penguins pictured below right. When pebbles are used penguins steal from each other's nests.

Sixty years ago a scientist painted some pebbles red and left them for penguins in a large colony to collect. This they did, and the pebbles started to travel as penguins stole them from neighbours' nests and added them to their own. The scientist then painted pebbles other colours, but the penguins clearly preferred red to any other colour

because the red pebbles travelled right across the rookery, as a penguin colony is called.

The interesting point about this experiment is that it was carried out on Adélie penguins which, apart from the little blue penguin, has less red than any other. It has only pale pink on its feet. It is almost as though Adélies, lacking red in themselves, had a mania for it in other things.

Magpie The European magpie was first known as a pie and was later named as Maggie pie or magpie, as we saw in Chapter I. Because a magpie has striking black and white plumage it soon became the fashion to speak of any animal boldly marked in this way as pied.

Later there came a change in usage and pied was used to denote any animal or garment that had patches of two colours. Strictly speaking, however, pied should be black and white only. Strangely, the word has tended to go out of use in modern times

except in the case of magpies and black-and-white horses, which are called piebald.

It often happens that a black-and-white pattern acts as a recognition signal, so that members of the species recognize each other readily. That might be the use of the pied plumage of the magpie, since it is a woodland species.

There are, however, other animals including the skunks, in which the piebald pattern serves as a warning that they are unpleasant for one reason or another. This could be true also for the magpie which is a robber of small birds' nests and a doughty fighter. Most members of the crow family to which the magpie belongs, can be aggressive, but a pair of magpies will attack again and again, in a most unpleasant manner.

So that pied plumage could have a dual purpose: a recognition sign for the magpie's friends and a warning signal for its potential victims.

Genet (Left) White shows up well at night even when there is only starlight and no moonlight. Nocturnal animals that have white anywhere on the body can be recognized by the pattern of their white patches. Moreover, white fur in these patterns seems to be almost luminescent. The African genet, a relative of mongooses, has white patches on its face and black and white rings on its tail. In the black of night the genet, seen head-on, shows just these white patches, which seem almost to glow. In side view, less is visible of the face patches but the white rings in the tail also show up well in the dark. So one obvious effect of the genet's coat pattern is to be easily recognizable by others of its kind.

To other animals, however, this pattern is far less conspicuous. The genet is a carnivore and one of the most agile and cunning of forest hunters. Slipping through the moonlight-dappled foliage in search of insects, birds and rodents, the elegantly striped and spotted genet is well concealed from its victims.

Skunk So far we have had to speculate on the value of the pied pattern to the animal possessing it. When we come to the skunks there is no doubt. There are eight kinds of skunk in America, four in North America. All have two things in common: they are pied and they are unpleasant. Their obnoxious quality is the readiness with which they discharge from their anal glands – the glands under the tail – a fluid with a most repulsive smell.

Their redeeming quality is that normally they give notice of their intention to do this. They do so first by their bold pattern of black and white. They do so also by the way they behave prior to discharging the fluid. They adopt an unusual or threatening attitude. This varies from species to species. Some stamp on the ground with the hind feet. The striped skunk turns its back on its target, lowers its head and raises high its bushy tail. The spotted skunk walks on its front feet with its hindquarters in the air, in a sort of handstand for about five seconds at a time.

Puffin (Below) The puffins belong to the auk family. Other members of the family are the guillemots or murres, razorbills, auklets and murrelets. All resemble penguins in having a white front and black back, but whereas penguins are confined to the southern hemisphere the auk family is confined to the northern hemisphere, The auks live round the rocky coasts. They swim and dive well, flying underwater to catch small fishes.

As a family, the auks break the monotony of their pied plumage even less than the penguins. The puffins are an exception. They have yellow feet and legs and a large bill that is coloured red, yellow and blue-grey in vertical stripes. Since the bill is parrot-like, puffins are sometimes called sea parrots. These colours are used in courtship. At the beginning of the breeding season the legs and feet become vermilion, a sheath grows over the bill and its colours intensify and become the more prominent by contrast with the white face. During courtship the male approaches the female with mincing steps and nibbles her bill. Then the two stand breast to breast, shake their heads rapidly to and fro and rattle their bills together.

Badger Badgers belong to the same family as skunks. Others in the family are weasels, otters, mink and polecat. All are able to discharge an unpleasant liquid from the anal glands, although none can equal the skunk at the game. Badgers are not pied in the usual sense but they do have prominent black and white marks on the head. It was natural therefore, judging by their near relatives, the skunks, to suppose that these marks were a warning coloration. That is now doubtful, although it cannot wholly be ruled out. What seems more certain is that these are recognition marks, for use between members of a species. Badgers are active almost entirely at night and a badger needs to know, when it hears a commotion in the undergrowth close by, whether it indicates the presence of friend or foe. The contrasting black and white on the head, as in this European badger, would give instant recognition.

The common badger of Europe also lives in parts of Asia. The American badger, however, is, despite its similar general appearance, a different species, and its head is not so prominently striped.

Piebald and skewbald As we have seen, the word 'piebald' was used in the Middle Ages for any animal that was black and white. Today it is used mainly to describe a horse patterned white with large black patches. Such a horse is also called a pinto. In the late seventeenth century we meet for the first time the name 'skewbald'. This was used for animals patterned like piebalds but in white and brown or white and red. As with piebald, so with skewbald, it is today mainly used to describe horses. Skew means out of the true, and presumably either brown and white horses suddenly appeared in the seventeenth century or, as is more likely, somebody suddenly realized there was no special name for this pattern and coined the word.

Pinto is a name derived from a Spanish word meaning 'painted' and the two-colour pattern can appear on many breeds of horse. But in the United States the pinto, either in its piebald or skewbald form, is virtually regarded as a breed in its own right and is an extremely popular riding horse.

Sacred ibis (Right) A striking bird with pied plumage is the sacred ibis of Africa. As in the case of other pied birds, there is also colour, but this is hidden except when the bird flies. Its body and tail are white, its head and neck and the long downcurved bill are black. In flight it shows iridescent green flight feathers and violet secondaries; the bare skin under the wings and at the sides of the breast is scarlet.

It is difficult to say what advantage the sacred ibis derives from its pied appearance, for it will adapt to many kinds of habitat. Although no longer found in Egypt, where it was once revered, it roams over much of the rest of Africa south of the Sahara. The sacred ibis feeds on insects, especially locusts and grasshoppers, as well as fishes, small crustaceans and molluscs, small reptiles, frogs and worms.

Whether or not the bold colouring had any effect on the bird's natural enemies, the sacred ibis certainly impressed the Ancient Egyptians. They regarded it as the incarnation of Thoth, their god of wisdom and learning.

Ostrich Only the male ostrich is pied. The female has a brown plumage in which the feathers have pale edges. There is a good reason for this.

The ostrich is the largest living bird, flightless but a good runner, nearly 8 ft high and up to 350 lb weight, able to exceed 40 mph with long strides of its long strong legs. These legs can also deal a very powerful kick. Normally, therefore, the ostrich has little fear from enemies, but at nesting time it is vulnerable, as are most other animals.

The task of incubating the eggs is shared between male and females. (There is some doubt as to whether ostriches are monogamous or polygamous. Modern opinion now seems to be that there is usually one male and three females to a nest). The male incubates the eggs at night when a black body shows up least. No doubt the white feathers in his wings and tail serve to break up the outline of his body, making him look less like a bird than would otherwise be the case. A female relieves the male in the morning and incubates for most of the day. During that time she holds her long neck down near the ground and with her pale-edged plumage takes on the appearance, certainly from a short distance, of a dry thorn bush in her semi-desert habitat.

Fallow deer Fallow deer of Europe are medium-sized deer with palmate antlers. That is, the antlers are flat blades with five or six finger-like projections along one side. The normal colour of the deer in summer is a deep fawn with prominent white spots on the flanks and black along the back and tail. The winter coat is a greyish fawn with little or no spotting. There are, however, more colour variations in the fallow than in any other species of deer, or any other wild mammal. The coat may be cream, sandy, silver grey, even white; and melanistic (black) individuals also occur.

Although fallow deer are native to western Europe and the Mediterranean countries they have been taken to several other parts of the world, including North and South America, Australia, Tasmania and New Zealand. They have been favourites as park cattle for at least the last thousand years, probably because of their varying colour. As usual, when there is such a colour range, owners of park herds often strive to have all-white or all-black herds.

Bittern An example of 'cryptic' coloration.

Camouflage techniques

The most complete method of hiding would be to become invisible. Some animals almost achieve invisibility by being transparent or nearly so. How they manage this is at present a mystery. All we know is that certain aquatic animals are transparent to the point of invisibility, except when they feed. Others come near to it, with only the skeleton showing in a kind of X-ray effect.

The next best thing to being invisible is to look like something else, by having the same shape and colour, as with the leaf butterfly on page 111.

Something very near to this kind of invisibility can be achieved by having a more or less uniform neutral colour, such as grey, or a colour that readily tones in with other colours, such as brown. A mixture of these two is commonly found in the coats of mammals and is sometimes called a pepper-and-salt mixture. This is when the hairs of the fur are brown, tipped with grey.

One thing cannot be emphasized too often when dealing with camouflage: it is effective only when the animal is still. Even then the disguise may be penetrated, as animals do not necessarily hunt by sight alone. Many mammals find their prey more by scent than sight. Birds which normally have a poor sense of smell, bring hearing to their aid. Reptiles and amphibians may use smell and hearing, but too little is known to be sure about this. More significantly, the structure of their eyes is such that they are acutely aware of any slight movement.

From this it follows that no camouflage gives total security. It helps an animal to escape from enemies and, more important, provides it with added safety when sleeping or resting – that is, when it is off-guard and at its most vulnerable.

Animals have to rely on their senses, of sight, smell or hearing, to discover their prey or their enemies. So prey animals, as well as those that must for other reasons avoid being seen, make use of disruptive lines or spots to break up their outline and make themsleves less conspicuous. The bittern, which is a member of the heron family, is a large bird and it looks terrifyingly large when about to attack, spreading its broad wings to their full extent. Its colour is brown – or, more accurately, brown with buff and dark brown to black markings. The dark markings of the throat form vertical lines. When the bittern is in its natural habitat, among reeds, and it stretches up its head, the buff neck with dark stripes tones in with the stems of the reeds and their shadows.

More remarkable still, the dark lines on the bittern's back and wings are vermiculate, or wavy. So whether a bittern is standing against a background of reeds, bushes or bare earth, crouched with its wings spread or stretching up its head, it seems to vanish into thin air. You can even set a bittern down on the greenest grass, then walk away a few paces, turn round, and have difficulty in seeing where it is. It is almost an optical illusion.

This kind of pattern is known as 'cryptic', from a Greek word meaning 'hidden' and a cryptic coloration is found in many animals, especially insects. It is a colour or pattern of colours broken up by stripes, bands, lines or spots. And while most animals do not use a full crypsis, many use lines and spots sparingly, to give just enough 'deceit of the eye' to make themselves reasonably safe.

Roe deer Although baby deer can stand on their feet within an hour or so of being born, it is some time before they can run swiftly. So the mother gives birth to her baby among dense undergrowth. There it lies while she goes away to feed, and she returns from time to time to suckle it. Meanwhile, the baby – in this picture a roe deer kid – is protected by its spotted coat. In these early days of its life the baby deer is very hard to see against its natural background, especially as its instinct is to lie quite still.

The adult roe deer is not spotted. Its coat is reddish brown (russet) but the important thing for the roe deer is its ability to hide. One can spend a whole day in a wood where 20 to 30 roe deer are known to be living and yet not see a sign of one of them. This is not due so much to their surroundings as to their ability to move quietly through the undergrowth and take advantage of every bit of cover it affords.

Burchell's or common zebra There has been much argument about the value to a zebra of its stripes. The usual view is that they make a zebra invisible but the fact remains that a group of zebras on the African plains stand

out with particular clarity. Yet a brown wildebeest standing nearby will be almost impossible to see at that distance. Perhaps the truth is that at certain times, as when there is a slight mist, or at twilight, the zebras suddenly seem to disappear. Another view is that the bold striping serves as a warning to lions, which are the main enemy of zebras. Certainly many a lion has suffered jaw injury from being kicked by a zebra. So perhaps the stripes signal a warning to the lion to approach with care.

There was one species of zebra known as the quagga that formerly lived in southern Africa. It was striped only on the front half of its body yet it seemed to suffer no disadvantage from this because until man came along with his firearms and wiped it out its herds numbered tens of thousands.

Puma It seems at first sight that there is a distinct difference between spots and stripes. Yet stripes can readily be broken into spots and spots can run together to form stripes. This can be seen in many animals. In the cat family spots and stripes are almost the rule, and a number of species possess both. In some of the larger members such as the lion, spots are present in the cubs and are lost in the course of growing up. The same is true of the cougar, or puma, of America. It is also known as the mountain lion, because when adult it looks like a lioness. When very young, however, it is spotted like a lion cub.

The colours of the adults show a wide range. The most common are brown and brownish red, but may be slate grey, bright red or yellowish red. The colour seems to have nothing to do with age, sex, the season of the year or the locality in which the animals are living. All-black pumas have been seen in South America, and also the occasional pure white albino.

Leopard cat Despite its name, the leopard cat has little to do with the leopard except that it also is a member of the cat family and has spots. It lives in the forests and grasslands of Southeast Asia as far east as the Philippines. Not much larger than a well-grown domestic cat, perhaps 4 ft long including a 2 ft tail, its spots are helpful in stalking prey. The habit of most cats is to stalk slowly, carefully, then kill with a final pounce. The spots break up the outline of the body, so the leopard cat can creep up unawares on its victim. The stripes on the face also help to hide the hunter but their primary use almost certainly is to enable leopard cats to recognize each other.

Stripes that form an intricate pattern on the face, as in the leopard cat and other members of the Felidae, may also help in communication. Cats and dogs express moods, especially hostility or menace, by varied facial expressions, these being effected by wrinkling the skin covering the face. The dark lines tend to accentuate the wrinkles, making the expression look more ferocious.

Raccoon Several night-hunting animals wear what has been called a 'highwayman's mask'. One of the best known is the raccoon of North America. It belongs to the same family as the giant panda and it is tempting to suggest that the black patch around each eye of the panda may perhaps be the remains of a highwayman's mask which, in its ancestors, was as complete as that of the raccoon. The black-footed ferret of North America has large black eyepatches that meet across the nose, forming a halfway stage between the panda's black eyes and the raccoon's mask.

An interesting point about the raccoon is that it also has white face patches, above and below the mask, and that its bushy tail has dark and light rings. As in the case of the African genet, the light patches and tail rings identify it to its fellow raccoons.

The raccoon is an inquisitive and intelligent animal. Its protective colouring is, nevertheless, of little value against its principal enemy, man, who has for centuries hunted and set traps for it, chiefly for its fur. This has not prevented the animal from venturing into city outskirts. When the lid of a dustbin suddenly clatters in the dead of night the chances are that the culprit is a raccoon, intent on turning over the garbage for something to eat.

Giraffes with impala This is a scene in East Africa with giraffes browsing the foliage of trees and impala ewes in the foreground. It shows the great advantage giraffes enjoy in being able to take food which smaller animals, such as impala, cannot reach. The great size of the giraffe would be a drawback, in making it readily seen by enemies, were its outline not broken up by the spots and blotches on its coat. On the open plain nothing would disguise such a large animal, unless it were far away, but among trees a giraffe is not easily seen.

The chief enemies of giraffes, especially of the young, are lions, but leopards, and even crocodiles, have been known to kill them. A determined lion can kill a full-grown giraffe but it must take the risk of a fatal kick from the intended victim's large hoofs and strong legs. On one occasion, a lion killed a giraffe in typical fashion by leaping onto its back and biting into its neck. But when the giraffe collapsed the lion was crushed to death under its one-ton weight.

Clown loach and tiger barb There are fishes in many parts of the world whose bodies are patterned with bands of black. This is especially true of the loaches and barbs, small freshwater fishes. Two kinds are seen here, both belonging to fresh waters of Indonesia. Those above are tiger barbs, and the two larger fishes below are clown loaches. Both types of banding are disruptive (breaking the natural contours of the body) and there is another point to note. The foremost band on the clown loach passes through its eye. In most animals the eye is a give-away. Some animals that are well camouflaged shut their eyes when at rest. In fishes, which cannot shut their eyes, the black band makes the eye less obvious.

There are many other fishes that have this black or dark band through the eye, and so have a number of frogs, reptiles, birds and mammals. Although it is dangerous to generalize, it does seem likely that it is an aid to camouflage.

Angelfish In the picture (right) are five kinds of tropical freshwater fishes. At the top are neon tetras, like neon signs, below and to their left are black neon tetras. To their right are two black widows. In the centre of the picture are two angelfishes and at the bottom, on the stones, an armoured catfish. All are very obvious in this situation, in an aquarium. But put them all back into their native waters and they would disappear as if by magic. The boldly marked angelfishes and the black widows disappear among water plants where their disruptive vertical stripes mingle with the shadows of the stems and slender leaves.

There are several species of angelfishes kept in aquaria. In the wild they are found in the Amazon and other rivers of South America, but today they can be bred in captivity and many of them are hybrids, obtained by crossbreeding the two most popular species. These two species are not easy to tell apart in their native rivers for both have a silvery, disc-shaped body with four prominent vertical black bands.

Greater kudu To anyone not accustomed to watching wild animals in their native habitats it may seem hard to believe that a few stripes can make much difference. But as this picture of a greater kudu shows, even modest striping may conceal an animal most effectively. This massive antelope from Africa, with its magnificent twisting horns, would not seem to be easily overlooked. Yet if the photographer had attempted to take its picture when it was farther back in the undergrowth, the kudu would have been very difficult to pick out from its background of thick vegetation. The coat of the greater kudu bull (pictured here) is greyish brown whereas that of the cow is a clearer brown. Both sexes have thin white stripes over the back and down the flanks, and white marks on the face, especially between the eyes and on the chin. The white coat stripes can be seen too on the newly born calf.

The greater kudu's close relative, the lesser kudu, being smaller, is not so skilful in defending itself against predators and therefore has even greater need of camouflage in the bush. Its striped coat pattern is similar to that of the greater kudu.

Siberian tiger Everyone takes it for granted that a tiger's striped coat makes it very hard to detect when the animal is crouched in long grass. Certainly tigers are adept at keeping out of sight, which is one reason why there are so few photographs of them taken in the wild. They live in many kinds of habitat, not always in long grass, but have a preference for dense forest. In open country and in daylight a tiger is quite conspicuous, but it hunts mainly by night. Then, in shadow or in moonlight, especially at dusk or at dawn, whether against a background of reeds or grass, or in dense jungle, it is difficult to see the outline of its body.

The tiger is usually assumed to live only in tropical lands and certainly it roams freely through India and Southeast Asia, including the islands of Sumatra and Java. But its range extends farther north as well, into Iran, Mongolia and even Siberia. In fact the tiger probably came originally from Siberia and was gradually forced southward as a result of the Ice Ages. The Siberian tiger is larger than its jungle relatives in the south and because it has to withstand the intense cold of high plains and mountains its fur is also longer. Furthermore, it is not so prominently striped and its fur is paler in colour.

Cheetah The cheetah lives in Africa. It used also to be found in southern Asia but has virtually been wiped out there. Known too as the hunting leopard, the animal was at one time captured young and trained to hunt the blackbuck, a very speedy antelope. For the cheetah is also a remarkably swift runner. It can accelerate to 45 mph in two seconds and reach speeds of 50 mph and more – some people say 60 or even 70 mph. But it is esentially a sprinter. If a cheetah does not overtake its quarry within a quarter of a mile, it stops and sits on its haunches and apparently loses interest.

The hunting leopard is a member of the cat family but is characterized by very long legs, rounded head and small ears. Its body is spotted, not as in the leopard in which the spots are arranged in rosettes. The light and dark markings of the coat are useful for concealment in the long grasses of the savannah. When the animal is crouched or squatting, especially in grass, it may not be easy to distinguish it from a leopard, but the distinctive black stripes, or 'tear marks', on the face give instant recognition.

Vulturine guineafowl Guineafowl are game-birds related to pheasants and, like them, feed on leaves, seeds, insects and other small animals scratched from the ground. The vulturine guineafowl is the most beautiful of them, with long white-shafted feathers extending from the neck over the back and breast. It lives in the dry areas of Central Africa, on the bush-veldt. Its spotted plumage makes it difficult to see, in spite of its bright colours, so long as it remains still. It goes into undergrowth to nest, where it is even more effectively hidden.

The half-dozen species of guineafowl are African but some have been taken to other parts of the world, to be used as poultry. They also serve as efficient guardians of poultry because they so readily give vocal warning of the approach of marauders, animal or human. All guineafowl have a spotted plumage, except two West African species. One is the black guineafowl, the other is the white-breasted, black with white patches. Both live in the dense, dark jungle.

Emu The emu ranges over the whole of Australia. Second only to the ostrich in size, it stands 5–6 ft high and weighs about 100 lb. It has wings but these are only tiny and hidden under the long double feathers that clothe the bird. For protection the adult emu relies on its strong legs for speed in running. The male takes over the eggs as soon as they are laid, in a nest on the ground, and has sole charge of the chicks. These are striped brown and white, in contrast to the adult plumage that is blackish brown. The stripes are disruptive, that is, they tend to break up the outline of the body, making the chicks hard to see in long grass or when they are still.

Despite its superficial resemblance to the ostrich, the emu belongs to a different family. Apart from its slightly smaller size, the emu has three toes on either foot whereas the ostrich has two. It is in fact related to another flightless Australasian bird, the cassowary.

The feathers of both these birds appear to be double. In most birds the feather is supported by a main stem or shaft and there is a very small aftershaft; but in the emu and cassowary the aftershaft is as long as the main shaft.

Wild boar The domestic pig or hog is descended from the Eurasian wild boar, which is a far more formidable beast. It is heavier than the average farmyard pig, more aggressive, with razor-edged tusks, and it is black or blackish. Furthermore, its coat of bristles is denser. Altogether, the wild boar is no mean opponent for any predator. Its piglets are in need of protection, which is given by the sow, in full measure if they are attacked. Their coat, instead of being a uniform colour, is brownish with light stripes, a combination that makes the piglets difficult to see in the scrub of the woodlands into which they are born, provided they remain still.

The wild boar extends across Europe and Asia and throughout this range there are many local races. From the study of bones on prehistoric sites it seems that each region throughout Eurasia domesticated its own local race but that these domesticated forms have been moved about and cross-bred since Neolithic times. Numerous pictures of pigs have been preserved, from cave paintings to modern drawings. Yet nowhere do we find young pigs portrayed showing stripes. Although it is unusual for juvenile characteristics to be lost completely in the course of domesticating a species, this is what must have happened in the case of the pig, for the stripes of the baby boar simply vanished in its domesticated descendants.

Gaboon viper The Gaboon viper is the largest of the puff adders of Africa. It is up to 6 ft long and has fangs that may measure as much as 2 ins in length. Despite its forbidding size, it is relatively harmless in that it seldom attacks people. This is just as well for not only is its venom very potent but it is a master of concealment. Its body is gaudily coloured yellow, purple and brown in a geometrical pattern, this being very obvious when the snake is laid out on a single-colour background. When it lies among dead leaves, however, which is its usual habit, it is hard to tell where the snake ends and the leaves begin.

Puff adders are so called because of the loud hissing noise they make when puffing air from their lungs. The African puff adders are highly venomous; but the quite harmless North American hog-nosed snake has also been called a puff adder because it blows air loudly, as if in warning, when disturbed.

South American leaf fish This aquarium contains South American guppies and, below them, a leaf fish, which is a predator, feeding on other fishes. The guppies probably have not seen it because the leaf fish looks remarkably like a dead leaf, especially when swimming. It swims sideways, very slowly, giving a most creditable imitation of a brown leaf being gently wafted along by the current.

Leaf fishes of various kinds are found in the Amazon and Rio Negro basins of South America and also in West Africa and South-east Asia. All resemble floating leaves and they are, with one exception, a mottled brown. Some have a short barbel on the snout that looks like a leaf stalk, especially when they float head-downwards. A leaf fish approaches its prey slowly until it is within reach. Then it puts on a spurt and opens its large mouth with extensible jaws. Some species of leaf fishes are capable of engulfing fishes half their own size.

Crab spider An animal likely to be attacked tries to hide. The crab spider reverses this. It needs to be hidden in order to attack the insects forming its food. A crab spider takes up position on a flower, usually at its centre. In this instance the flower is a milkweed. There the spider remains very still and slowly its colour changes by hormone action to match that of the flower, taking two to three days to do so. When the change is complete, the human eye has difficulty in seeing the spider. A butterfly's compound eyes are even less efficient, but better equipped for detecting movement. So long as the spider remains still there is a good chance of an insect landing on it, to be captured.

An experiment was devised to determine what effect the yellow colour of the spider had on insects visiting the flower where it was lying in wait. It was noticeable that the majority were strongly attracted to yellow flowers in which yellow pebbles the size of the spider had been placed, as against yellow flowers with black pebbles of a similar size in the centre.

Stick insects (Left) Stick insects, known in North America as walking sticks, have a long, slender body, thin legs and delicate antennae. They are usually coloured green or brown and when they draw their long legs into the body and hold their antennae rigidly out in front, they look remarkably like twigs or sticks. They add to this deceptive appearance, thereby duping insect-eating birds and mammals, by remaining absolutely motionless, even when they fall to the ground among dead leaves and stalks. The rounded brown eggs, looking like seeds, drop naturally to the ground as they are being laid.

The largest stick insects grow up to 9 ins, but those that are frequently kept in schools and laboratories are seldom more than 4 ins long. One species from Queensland, Australia is a stick insect with a difference. When very young it looks like an ant as it runs swiftly over a plant. Then it begins to slow down. Its body and legs begin to flatten out, so that it looks less like a stick than a leaf. When preparing to move, it starts swaying gently from side to side, as if it were trying to mask its telltale movements by assuming an even more leaf-like appearance.

Leaf butterfly (Top right) One of the best known examples of animal camouflage is the Indian leaf butterfly. When the butterfly lands on a plant and closes its wings, it looks exactly like a dead leaf. The shape is identical even to a small 'tail' at the rear of the closed hindwings, that resembles the leaf stalk. The undersides of the wings are brown and have markings that look like the midrib and veins of a leaf. The wings may even have holes in them, like torn dead leaves, or markings that make them look like fungus-infected leaves. If disturbed, the butterfly flies swiftly away, flashing the orange or blue on the upper surfaces of the wings. Then it settles on the ground, closes its wings and becomes part of the leaf litter, so performing a marvellous vanishing trick.

The orange and blue on the upper wing surfaces are called flash colours. Many butterflies and moths have these. They flash as the insect flies, making it difficult for a predator to follow its line of flight. Then, as it settles, the flashing colour disappears and the pursuer is likely once more to be confused.

Bark mantis (Below left) As we saw in Chapter 5, praying mantises are insects that are living insect traps. Their first pair of legs are long and armed with spines. The mantis sits more or less erect with its front legs folded, almost as if in prayer. When an insect alights near it the gin-trap legs shoot out and jack-knife over it, catching and holding it as in a gin-trap. From this it is clear that a praying mantis is an animal that must hide to attack. It must not be readily seen if it is not to go hungry. This bark mantis of Africa looks like the bark on which it habitually lives.

Oak beauty (Right) A common moth in the oak woodlands of Europe is the oak beauty,

seen on the wing early in the year when the trees are bare of leaves. Normally it rests by day on branches and trunks which are covered with a thin layer of lichens and mosses. The wings of the moth are coloured and marked in such a way as to match almost perfectly the lichen-covered bark.

One of the basic differences between a butterfly and a moth is that moths are usually active by night and rest by day. Butterflies do the reverse, remaining out in the sun all day and retiring to rest at night. Also, when butterflies come to rest they normally hold their wings close together over their backs. Moths, on the other hand, usually rest with the wings spread. So any camouflage a butterfly possesses is more likely to be on the undersurfaces of the wings, while moths more often have the upper surfaces camouflaged.

Moths can be seen taking wing in large numbers at dusk, yet during the day they are very hard to find, so well are they camouflaged.

Rabbit (Right) Rabbits were domesticated from a very early period, precisely when is not known. A few of the strains produced under domestication are the Angora, an albino (pure white with pink eyes), noted for the length and fineness of its fur, the lop-eared rabbit, with drooping ears that may reach the ground, and the Himalayan rabbit, pure white except for the nose, ears, tail and feet, which are black.

Physical features and colours not normally found in the wild are often the result of selective breeding by man, as is the case with these and other types of domesticated rabbit. But such changes may also come about as a consequence of long isolation, as happens when wild mammals find their way from the mainland to an island, either by natural means or having been taken there by man. Under natural conditions such animals may begin showing changes in size, weight, body shape and fur colour in as little as 70 years after arrival. Rabbits were being kept on islands or in warrens at least a thousand years ago. They were isolated in this manner because it was realized that they could be highly destructive to crops and pastures. Doubtless rabbits kept in isolation or in captivity soon began to show changes in coat colour, length of ears and weight of body; and in due course man lent a hand, intensifying these changes.

Kangaroo (Below) The kangaroos and wallabies are the best known examples of a marsupial or pouched mammal. Born at a very early stage of development, when only bean-sized, the baby kangaroo makes its way to the mother's pouch (marsupium) and remains there, fixed to one of her teats, for about three months. Even when it is too large for suckling it seeks warmth and shelter in the pouch.

There are many species of wallaby and kangaroo and although the majority are found in the dry deserts of Australia's interior, some are inhabitants of woodland and forest. There are wallabies with black bands on the body and with black rings on the tail, but as a rule the coat colour is plain and unremarkable, often described as 'pepper-and-salt'. This term was originally applied to a type of cloth made of dark and light coloured wools woven together, showing a pattern of small, intermingled spots. Animal fur which contains the same blend of light and dark hairs has a neutral quality that provides reasonably effective concealment for its owner against a variety of backgrounds.

Blacksmith plover (Top right) These long-legged, long-billed birds are of the kind known as waders. They live near the sea or on the edges of large lakes, where the ground is sandy. The plovers make a scrape in the sand into which the female lays her eggs, which are spotted and so like pebbles that it is almost impossible to spot them without diligent search. The chicks can run about within a few minutes of hatching. When danger threatens they freeze, crouching motionless on the sand at the alarm call from the parent. Their mottled downy plumage harmonizes perfectly with the ground, as can be clearly seen with these chicks of the blacksmith plover, seen here on the volcanic sand near Lake Nakuru, in Kenya.

The blacksmith plover has a call that sounds like two pieces of metal being tapped together. It is a largish bird, 11 ins long, black, white and grey in strongly contrasting patches. When sitting on its nest on a stretch of the sand, it can be seen a mile away, but because it has a long flight distance (the distance it allows anyone to approach before it flies off), it is very difficult to locate either the nest or the chicks.

Koala The koala, popularly known as the Australian 'teddy-bear', is also a marsupial. But whereas the pouch of the female kangaroo opens towards the front, that of the female koala opens to the rear. Since the animal is a tree-dweller there does not seem to be any obvious advantage to this arrangement and the baby has to hold tight in order to avoid falling out, particularly when the mother is clambering up a trunk. When it grows large enough, however, it rides pick-a-back on the mother.

The name 'koala' is an Australian aboriginal word which is said to mean 'no water'. It used to be assumed that koalas never drank, but they have, in fact, been seen drinking from pools left after rain. Furthermore, the koala is a good swimmer and after crossing a river will lick and swallow the water collected in its fur.

Weighing up to 20 lb, the koala lives exclusively in eucalyptus trees and feeds on leaves and shoots, rarely descending to ground level. The bark and leaves of these trees are grey and this colour is matched by the koala's ash-grey fur.

Phantom or **glass larvae** (Left) There is one gnat in Europe which does not bite or suck blood. Apart from this peculiarity it has a special interest for naturalists because its larvae, known as phantom or glass larvae, are almost completely transparent (except for the eyes) and nearly invisible so long as they do not move. They are able to remain suspended motionless in water with the aid of two gas-filled floats on the body which can be inflated or deflated as the density and pressure of the water vary. The advantage to the larvae of remaining still is that the water insects and tiny crustaceans which make up their food swim up to them without suspecting their presence. The larvae then seize the victims with antennae which are armed with curved spines.

It may be that these phantom larvae hold the real clue to the value of transparency in animals. They become visible momentarily as they flick their body to dart away through the water, and become almost invisible once more when they stop moving and stay still. This produces something akin to the flash effect already noted in the case of the leaf butterfly.

Sardines (Left) A cylinder lighted from above will have a shadow along its lower side, showing as a dark patch. If the underside is painted a lighter colour the effect of the shadow is offset and the contour of the object is less conspicuous. This can also be applied to animals, when it is known as counter-shading, the effect being more common in water animals than land animals.

Sardines are small fishes that live in huge shoals, and this should alone give them some measure of protection against predators. They are, however, countershaded. Out of water the back of each sardine is greenish, the flanks golden and the belly silver. Underwater, these colours look different, but the silver of the belly is prominent. Many other fishes also have silvery underparts and it is commonly assumed that this is effective camouflage when the fish is viewed from below against the light at the surface of the sea. That is to say, any predator swimming underneath will stand little chance of seeing its prey because the silver belly blends so well with the light of the sky beyond the water surface. Now that it is possible, however, for observers to go down and watch fishes 'at home', it seems that the pale belly camouflage is effective only until the fish becomes silhouetted against the bright circle of light emanating from the surface.

Glass catfish The glass catfish of the fresh waters of Southeast Asia is about 4 ins long. Its colour is yellowish but its body is as transparent as glass. Along the top of the head and back and on the edges of the fins there is a little black, so that the body is outlined. What advantage or disadvantage to the catfish this transparency may be is not certain. In any event, there seems to be a further attempt at disguise when the fish rests. It positions itself in mid-water with its body obliquely upwards and the lower lobe of the tail pointing downwards.

There is another transparent fish, the glass-fish *Chanda ranga*, living in rivers and estuaries

of India, Burma and Thailand. Every bone in its skeleton can be seen, except in the head, which is opaque. Just behind its golden eye is a silvery patch. This marks the position of the abdominal cavity, and it is suggested that the silvery covering prevents the growth of algae in the gut. There must be some benefit in this transparent quality, since in some parts of India the 3-in. fish is so common that it is netted in large numbers and used as fertilizer.

Apollo rasboras (Above) Apollo rasboras are freshwater fishes of the carp family living in the rivers of South-east Asia. They swim in shoals and as often happens with shoaling fishes they have the advantage of countershading and dazzle effect. Countershading, produced by the dark back and silver belly is probably of most use when

they become separated from the shoal. The dazzle effect is produced by the strongly reflecting silvery scales of the belly as the fish suddenly twists. As the many members in a shoal do this the effect on the predator seeking to snatch one must be similar to the dazzle of many blinking, twinkling lights. Dazzle effect and the flash colours of the leaf butterfly are therefore very similar.

Rasboras are popular aquarium fishes, slim of body with a forked tail. Individually their colours are not brilliant but in shoals they make an attractive display. In most species the countershading effect noticeable in the Apollo rasboras is equally striking, the belly either being silver or markedly paler in colour than the back and flanks.

Little egret and gulls (Above right) Examples of countershading among birds can be seen in

certain gulls. The egret in the foreground of this picture is white all over and since its body is raised on long legs the shadow effect caused by lighting from above must be negligible. In fact, the bird has no need of camouflage on the ground since it is most vulnerable to its main enemies – birds of prey – when it is on the wing. The gulls in the background of the picture have a pearl-grey back and white underside, offsetting the shadow effect. It is not clear why gulls should have need of such protective countershading. They are strong fliers, have a powerful beak and will flock together in the face of a common threat. Nevertheless, in many species it is very striking.

In this connection it is worth mentioning the herring gull and the lesser blackbacked gull. The herring gull has a pearl-grey back and flesh-coloured wings and legs. The

blackbacked gull is black on the back and wings and the legs are yellow. The two species encircle the globe in Europe, northern Asia and northern North America but they both differ slightly in their colours from one place to another. Over northwest Europe they are so different that there is no difficulty in recognizing them as two distinct species. Yet they form what is called a ring species. Apparently in prehistoric times there was one type of gull in eastern Siberia. The birds spread eastwards over North America and due westwards over Asia and Europe, their colours changing as they extended their range. Finally, the two groups met somewhere over the countries of northwest Europe, one appearing as the herring gull and the other as the lesser blackbacked gull. Since both have survived equally successfully, the only deduction is that it makes no difference to

this ring species whether its members are pearl-grey or black.

Gerbil At night, even in bright moonlight, the human eye is at a disadvantage. The eyes of animals that habitually hunt by night are far better adapted to darkness; and animals such as this small rodent from central Asia, the gerbil, also known as the clawed jird or sand rat, need protection from their keen-eyed enemies. Its coat is sandy and the underside of the body is white – an instance in which countershading is probably of considerable importance in offsetting the effect of the shadow cast by its own body. The camera registers more than the human eye and this picture perhaps gives an idea of how much a nocturnal hunter might be able to see.

Spotted salamander The spots of the spotted salamander of Europe are not intended to hide it. On the contrary, they make it more conspicuous, advertising that the salamander will not make a palatable meal. All salamanders have poison glands in the skin. In some, these glands may be small but they are effective. The first scientist to test this was a woman, who placed a salamander between her teeth and gently bit it. She found that in a short while she was suffering from a severe headache, although she had not taken enough of the poision into the mouth to taste it; and it is noticeable that if a dog, for example, takes up a salamander in its mouth, it will show obvious signs of distress. The spotted salamander has more of these poison glands than most other salamanders and lets the world know about it by carrying the yellow and black markings.

There is much confusion in the use of the names salamander and newt. Both mean lizard-like tailed amphibians. 'Salamander' is from the Latin and was the name given originally to the spotted species of tailed amphibian shown here. 'Newt' is a medieval English word for the related species living in Britain. Today the names tend to be used indiscriminately, and, in effect, they both refer to the same kind of animal.

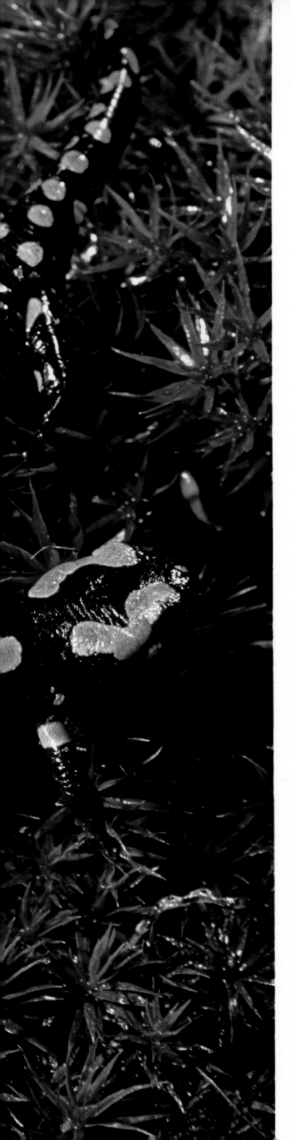

Warning colours

Many animals, as we saw in the previous chapter, need to find some way of disguising themselves in order to survive – by means of distinctive coloration, body shape, behaviour, or by a combination of these features. A few animals, however, achieve the same end by making no attempt to hide and by deliberately flaunting their conspicuous colours. In certain cases, the animals themselves may be in some way unpleasant or literally distasteful to a predator determined to attack and eat them. Their colours are said to be warning colours and their additional protective devices may be obnoxious fluids (emitted when the animals are touched or alarmed), painful stings, venomous fangs or poisonous spines.

It is not possible to lay down hard-and-fast rules, because some unpleasant and objectionable animals are also well camouflaged. Examples of these are the insects known as shieldbugs or stinkbugs, which have flattened, triangular bodies and are, to the human taste at least, highly obnoxious. But being coloured green, they are hard to see on the leaves of bushes they frequent, so watch out if you eat blackberries as you pick them. There are venomous snakes that look so like the dead leaves they lie in as to be virtually invisible. The most poisonous animal in the world is probably the stonefish of the Australian seas. This has a poison spine on its back and anyone who treads on it with bare feet will suffer agonies, if not death. Yet the stonefish is almost perfectly camouflaged on the seabed.

Provided it is clear, therefore, that there are these exceptions, we can proceed with a discussion of those animals that do possess warning colours. These are usually black, red or bright yellow, most often two or all in combination. Generally the animals tend to be either bright red all over, black and red, or black and yellow. When more than one colour is present, these are usually in bands, stripes or dots, and the colours themselves contrast strongly. In other words, the colours force themselves on the notice of anything that has eyes to see. The familiar wasp is a good example.

Another feature of animals with warning colours is that they tend to stay in the open and take up position in exposed places. The cardinal beetle, shown on page 16 is an example. Bright red, it walks slowly over the tops of grasses in full view of all passers-by. When it takes wing, it flies slowly. In all its movements it seems unhurried, as if determined to be seen.

The theory of warning colours is as follows: A predator, such as an insect-eating bird, picks up a cardinal beetle in its beak and quickly drops it, having tasted something unpleasant. The next time the bird sees a cardinal beetle it will leave it alone or receive a second lesson. Had the beetle been good to eat the bird would have looked for more, continuing its search until it had disposed of a dozen or more. So even if the first beetle happens to be killed by being taken into the beak, it will perhaps have saved the lives of many other cardinal beetles.

Predatory animals have to learn about warning colours. They do not instinctively recognize the full danger behind them. Therefore, warning colours need to be as striking as possible if only because they must be remembered. Tests have shown that the hunters readily learn, either at the first bad experience or soon after. Moreover, further tests have demonstrated that they usually remember the experience and even though they may be hungry refuse to be caught again by an animal with the same warning colours.

Interestingly enough, animals carrying warning colours are usually hard to kill. Insects, for example, will survive a pinch that would almost certainly end the life of a normal insect. So it seems that even the warningly coloured animals unfortunate enough to encounter inexperienced predators do not necessarily succumb to their attacks.

Banded coral shrimp (Left) The banded coral shrimps live by feeding on the parasites that infest the skin of fishes, so that they actually perform a service for these fishes by cleaning them. And since they do not move around but always stay in one place their 'customers' have to come to them. The prominent red and white bands on their body, legs and antennae advertise their whereabouts as unmistakably as the similarly coloured bands of a barber's pole in olden days. The two banded coral shrimps in this picture are stationed near some fanworms.

Sea slug (Top right) Land slugs are not very attractive. Their colours are usually very dull. This contrasts strongly with sea slugs, many of which are not only brightly coloured but exquisitely beautiful, especially those living in warm seas. Yet the way of life of both kinds of slugs is much the same. There has been little study made of sea slugs so we do not know how far, if at all, their colouring gives them camouflage when they are moving among brightly coloured sea anemones, or whether it is intended to have a warning effect.

Many sea slugs feed on tiny polyp animals known as hydroids, or sea-firs. These are related to sea-anemones and, like them, have large numbers of stinging cells in their skin. It seems that when a sea slug feeds on a sea-fir the stinging cells are swallowed with the rest of the tissues but are not digested. They migrate through the wall of the sea slug's stomach and reach the skin of the back. There they take up position and will sting anything trying to eat the sea slug. This is a most remarkable process about which little is known for certain, but if it does work in this way, it seems likely that sea slugs do possess true warning colours.

Pufferfish and scat (Right) In the estuaries of Southeast Asia from Thailand to Borneo, lives a pufferfish. Most pufferfishes live in the sea and are well known for their bright colours, for their habit of swallowing water to blow themselves up like balloons when danger threatens, and for having scales modified to spines which stick out when the fish is inflated. The trick of inflating the body is presumed to be a protection against being swallowed, but sharks are known to eat them. Another feature of pufferfishes is that some of them are poisonous, sometimes with lethal effect. Altogether, pufferfishes have unpleasant qualities and their bright colours and spots almost certainly serve as signals to other animals to leave them alone.

In the same habitat in Southeast Asia lives a scavenger called the scat or argus fish. The latter name refers to the polka-dot spots, and is derived from the mythical Greek hero with a hundred eyes. The former name comes from *Scatophagus*, the scientific name of the fish, meaning 'dung-eater', and refers to its habit of feeding on excrement and on almost any kind of refuse. The body is well-armed with spines, on the fins and on the gill covers. A reasonable interpretation would be that the bold spots are warning signals to other animals that scats are spiny and should be left alone.

These two unpleasant fishes sometimes strike up a partnership. The pufferfish has strong teeth. In the marine species these are used for chewing coral. In the estuarine pufferfish they are used for biting up large food. The scat, on the other hand, has a very soft mouth, and because it is incapable of coping with food of any considerable bulk will follow in the trail of the pufferfish, picking up any fragments of food that the latter happens to let fall.

Cinnabar moth One of the best examples of warning coloration is found in the cinnabar moth of Europe. The moth itself has blackish-green forewings marked with scarlet and hindwings that are scarlet. Moreover, the moth flies slowly, as if aware it has nothing to fear and therefore need not use energy on unnecessary speed. The caterpillar of the cinnabar moth is equally conspicuous. It is banded black and yellow, making it almost certain that it has distasteful fluids in its body. The caterpillars live in groups feeding on ragwort, eating the plant down to bare stems.

There is added protection from being gregarious (i.e. living in groups). If a predator, such as an insect-eating bird, tastes one of the caterpillars and finds it unpleasant, it will be less likely to try again, because it will associate the experience with the whole group. Also, if the production of a distasteful fluid involves an unpleasant smell, this will be intensified if the caterpillars live together in close company.

Wasp beetle The wasp beetle, of Europe, is one of the longhorn beetles, so called from their long 'horns' or antennae. It runs about on flowers with quick jerky movements, as a wasp might do, and its body, like that of a wasp is dark brown with bright yellow markings. Put a wasp beetle and a wasp side by side, however, and it soon becomes clear that they are two different insects. Yet anyone not too familiar with insects and their habits will hesitate to touch the completely harmless wasp beetle, simply because of its brightly striped colours. The same goes for insect-eating birds, so that for the beetle these colours are protective.

There are other examples in nature of harmless animals that 'imitate' dangerous animals, to which they are quite unrelated, in possessing similar warning colours and patterns. This kind of mimicry is not uncommon among moths and butterflies and is also found among reptiles, as in the case of the innocuous milk snake whose colours mimic those of the highly venomous coral snakes.

Burnet moth There are few more decorative moths than the Provence burnet moth of southern Europe. Its beauty, however, signals an unmistakable warning to birds to leave it alone. The moment it is touched it discharges a powerful poison, in the form of a yellow fluid, from the region between the head and the thorax. The poison is prussic acid and even in very small amounts leaves a lasting impression on any bird that tastes it. So a bird that makes the mistake of taking a burnet moth into its beak will release it as quickly as possible; and for the rest of its life it will carefully avoid touching any other moth bearing the same colours.

Carrion beetle One of the most tantalizing things about insects is that so frequently, when photographed, their colour looks different from reality. A black water beetle photographed underwater looks blue: and this bluish-looking carrion beetle is in fact black all over except for the last few joints of the antennae – an appropriate colour for an animal that feeds on the flesh of dead animals.

This beetle may be seen along river banks or on the seashore, both of them places where carrion is likely to be washed up and stranded. Sometimes another carrion eater gets in first as when blowflies lay their eggs on the carcase and the maggots hatching from them start to feed on the dead flesh. The carrion beetle is not squeamish: it will eat the maggots as well as the carrion.

To the human mind these habits are unsavoury but there is a sequel to the story. Such beetles are often infested with mites, seen here in brown clusters, small and distant relatives of spiders. They are said to be parasitic but some people maintain they help to clean the insect's body surface.

Blue-spotted (ringed) octopus Australian seas swarm with sharks and these are justifiably feared. Even more of a menace, however, as shown in recent years, is a small reef-living octopus, known as the blue-spotted or ringed octopus. Its body is only an inch long and this is spotted and banded in brown, and its eight arms are ringed in brown. At the centre of each brown patch is a blue spot, making this tiny octopus most attractive to look at but not to handle. In a year more bathers are likely to die from bites of the diminutive animals than from attacks by sharks.

It is unusual to find an octopus that does not, by the use of changing colours, try to hide. A century ago a scientist watched, under a low-powered microscope, unhatched baby octopuses, still in their transparent envelope. He saw the orange-brown colour in the pigment cells in their skin flashing, dying out and reappearing in another place, like sparks in tinder. To his astonishment, because the babies were still protected by the envelope, he discovered that they changed colour according to the background. Against a white paper they went pallid, against brown they went dark and when subjected to slight pressure they went red with irritation, just as happens with adult octopuses. This demonstrates that colour changes in animals are automatic from the very start.

Blister beetle (Right) Blister beetles give out a chemical, especially from their wing-covers, that causes blistering of the skin. Years ago, when physicians bled their patients in the hope of curing them, they often blistered them as well, using dried blister beetles for the purpose – a crude remedy that was none too soon abandoned.

Blister beetles are found throughout the world, especially in warm, dry climates. Like the one pictured here, they all carry warning colours in black, red and/or yellow. Usually the body is black with bold spots of yellow or red but it may be black with a small amount of yellow or red edging to the wing cases. The adults feed on plants and can be pests when they choose cultivated crops. The larvae feed on bee larvae, as described for the oil beetle on page 127, or on other insects. In North America larvae of some blister beetles feed on the eggs of locusts. These are laid in capsules in the ground. The blister beetle larvae actively search for the capsules, bite their way in and then lead an inactive life surrounded by an abundance of food.

Adder (Over) The adder or viper is a snake that ranges across Europe and northern Asia to the island of Sakhalin. It is not large by snake standards, rarely exceeding 2 ft in length. Its food consists of frogs and newts, lizards, small rodents, insects, slugs, worms and also small birds and their eggs. It is venomous but not dangerous compared with snakes in other parts of the world; and because poisonous snakes are rare in the regions where it lives, it can easily be identified by two prominent distinguishing marks. One is a black zigzag line running the length of the body. The other is a V-shaped black mark on the head.

In a sense, these markings can be classified as warning patterns, but the ground colour of the adder varies widely. Cream, yellow, silvery white, pale grey or olive adders are males. Red, reddish brown or golden adders are females. All have the black V and zigzag. Melanism is common, to a greater or lesser degree, and black adders occur very commonly in some localities.

African monarch There are fourteen species of monarch butterfly in Africa. They are sometimes called milkweed butterflies, their caterpillars feeding on plants known as milkweeds. The African monarch measures over 3 ins across the spread wings, which are orange-brown with black margins. The body, especially the front part, is ornamented with white spots. These are warning signals to birds and lizards that the butterfly gives out an acrid fluid if touched.

Monarch butterflies live in the tropics of both the Old and New Worlds. One of the few exceptions is the large showy monarch butterfly of North America, noted for its migrations. All monarchs have bold patterns of warning colours. They fly slowly making no attempt to hide and behaving in every way as if aware of their immunity from attack. The poison they carry is imbibed by their caterpillars from the more or less poisonous juices of milkweeds. Apart from the fact that they exude a noxious fluid, the

bodies of monarchs are tough and rubbery, able to survive a hard pinch that would kill most other butterflies.

Oil beetle Oil beetles lead a precarious life. The majority of oil beetle larvae hatching from a batch of eggs are doomed to die. The larvae climb onto plants and each tries to seize hold of the first hairy insect to alight. If that insect is a fly the larva must perish. Only if it is a bee, and the right kind of bee, is it carried to a bees' nest. Its first meal is a bee egg. Then it feeds on the bee larvae.

With so low a rate of survival it is not surprising to find that the adult beetles have a trump card against attack. When disturbed, as when picked up by a bird or by human fingers, a yellowish oily liquid oozes from its joints. This causes stains and gives out a most unpleasant smell. The liquid is in fact the insect's blood and the way it is discharged is called reflex bleeding.

Ladybird (See also page 54.) If there were no check on greenfly, scale insects and other plant lice, the world's vegetation would be in a poor way, so would our crops. They can multiply at a phenomenal rate and they spend their time sucking the sap from leaves and green stems. There are several insects that do more than anything else to keep the numbers of plant lice down. Foremost is the ladybird. The adult feeds voraciously on plant lice, and larva even more so. Here is a lunate ladybird of the grasslands of South Africa feeding on greenfly.

Wasp An out-of-doors meal or a picnic on a sunny day can be ruined the moment a wasp settles on the tablecloth. The sting can, of course, be painful but the panic effect that even a single wasp can cause among some people seems quite out of proportion to the real damage it is capable of inflicting. It may be due to the menacing buzz of the wings, the distinctive black and yellow jacket, the awareness of its ability to sting, or a combination of all these things. Insect-eating animals experience the same feelings of terror and for them the warning signal of the black and yellow bands is sufficient reason for giving the wasp a wide berth. Yet there are a few animals, of which the bee-eater (page 42) is one, that go to the opposite extreme. The bee-eater will deliberately chase and eat a wasp, knowing exactly how to handle its victim. Having snapped the wasp up in its beak it flies back to its perch and beats out the sting by hammering it against a solid surface.

Acknowledgments

The publishers would like to thank the following individuals and organizations for their kind permission to reproduce the photographs in this book:

AFA Colour Library: E. H. Herbert – 12 above, 93 centre; Ardea Photographics Ltd: F. Collett – 124 below, K. Fink – 21, W. Taylor – 42 left; Douglass Baglin: 112; Bille: 12 below; John Carnemolla: 11; Bruce Coleman: 80 below, 86 below, 88 D. & J. Bartlett – 22 below, J. Burton – 6, 13, 14 left, 16–17, 20 above, 23, 25 below, 26–29, 30 above left, 30–31, 32, 33 below, 35, 36 above, 37–39, 40 below left, 41 above, 42–43, 44 above left, 46, 47, 49, 50–51 centre, 50–51 below, 51 right, 52–61, 63 below, 64–65, 67 right, 68, 70, 71 above, 73 below, 75, 79 below, 81–85, 86 above, 87, 89, 90 below 91, 93 below, 94 above, 94 centre, 95–100, 101 below, 102 below, 103–105, 108 below, 109–110, 113 above left, 113 above right, 114 above, 115–123, 125, 126 below, L. Dawson – 79 above, S. Gallsater – 78, L. Lee Rue – 40 above left, J. Markham – 25 above, 40–41 below, R. K. Murton – 1, N. Myers – 14–15, 92–93, C. Ott – 18, 76–77, 80 above, J. M. Pearson – 22 above, G. D. Plage – 8 above, V. Serventy – 20 below, J. Simon – 9 above, N. Tomalin – 69, D. & K. Urry – 94 below, J. van Wormer – 19 above; James Cox: 62 below; P. M. David: 34 above, 50–51 above; Walter Deas: 34 below, 74 above; Robert Goodden, Worldwide Butterflies Ltd: 19 below, 44 below, 45 below, 48, 62 above, 72 below, 74 below, 111 above; Jacana Agence de Presse: 102 above; Frank Lane: 101 above; NHPA: F. Baillie – 44 above right, 72 above left, A Bannister – 33 above, 71 below, 73 above, 127 centre, J. B. Blossom – 66–67 below, 108 above, N. A. Callow – 127 below, S. Dalton – 90 above, 126 above, E. Elms – 63 above, P. Johnson – 93 above, P. Scott – 72 above right, T. Stack – 66 above, M. W. F. Tweedie – 45 centre, 66 below left, 111 below, 123 right, 124 above, 124 centre, 127 above; Okapia: 36 below; Photo Aquatics: H. Gruhl – 27 right, R. Lubbock – 114 below; Photographic Library of Australia: 2–3, 24, 113 below; Seaphot: C. Petron – 4–5; Tony Stone Associates: 30 below left, 107; S. A. Thompson: 8–9 below; Zefa/Pictor: 106